THE
RELAY RACE
TO INFINITY

Developments in Mathematics
from Euclid to Fermat

THE
RELAY RACE
TO INFINITY

Developments in Mathematics
from Euclid to Fermat

Derek Holton
University of Otago, New Zealand
University of Melbourne, Australia

John Stillwell
University of San Francisco, USA

World Scientific

NEW JERSEY · LONDON · SINGAPORE · BEIJING · SHANGHAI · HONG KONG · TAIPEI · CHENNAI

Published by

World Scientific Publishing Co. Pte. Ltd.

5 Toh Tuck Link, Singapore 596224

USA office: 27 Warren Street, Suite 401-402, Hackensack, NJ 07601

UK office: 57 Shelton Street, Covent Garden, London WC2H 9HE

Library of Congress Control Number: 2024040472

British Library Cataloguing-in-Publication Data
A catalogue record for this book is available from the British Library.

THE RELAY RACE TO INFINITY
Developments in Mathematics from Euclid to Fermat

ISBN 978-981-12-9632-1 (hardcover)
ISBN 978-981-12-9758-8 (paperback)
ISBN 978-981-12-9633-8 (ebook for institutions)
ISBN 978-981-12-9634-5 (ebook for individuals)

For any available supplementary material, please visit
https://www.worldscientific.com/worldscibooks/10.1142/13935#t=suppl

About the Authors

Derek Holton is professorial fellow at the University of Melbourne and emeritus professor at the University of Otago, New Zealand. He lectured at Melbourne University for 25 years and at Otago for 23. His research areas are in Graph theory/Combinatorics and Mathematics Education (largely in the area of problem solving). His two previous books with World Scientific Publishing are *A First Step to Mathematical Olympiad Problems* and *A Second Step to Mathematical Olympiad Problems*.

John Stillwell is emeritus professor of mathematics at the University of San Francisco, where he taught for 20 years after 30 years at Monash University in Australia. He is known for his expository writing in mathematics, for which he received the Chauvenet Prize of the Mathematical Association of America in 2005. Among his best known books are *Mathematics and Its History* and *The Story of Proof*.

Contents

Prologue

The Contents

I'm not quite sure what monster I have created. It all started by my thinking that not many adults or school students – or even academics for that matter – know much about the history of mathematics. You probably know the big names of the past and some of what they did, but you may not know what was happening around them.

It didn't take very long to realise that there were large numbers of these outstanding historical mathematicians and I had to cut the number down or I would be writing an encyclopedia. At that point, I just became selfish and I added names that I thought were good to list and then I found names that I had never heard of before and then I was picking up professional mathematicians from all over the place.

What I was trying to do here was to introduce you to some prominent mathematicians and, where possible, say something about them, their times, where they lived, what they ate, and who their friends were. I also hoped to show you some of the mathematical ideas they had and where this led mathematics. In addition, I hope to provide some mathematics for you to think about that is related to their work. Overall I was trying to produce a concise history of mathematicians and mathematics that you might find interesting.

To make my task reasonable, I decided to start off by looking at Euclid, Liu Hui, Fibonacci, Fermat, Euler, Gauss, Noether, and Turing. But before I knew it, I had 300 pages for the first four people and their friends and acquaintances. I decided that eight were too many for one book. So I've restricted this book to those first four important people, Euclid, Liu Hui, Fibonacci and Fermat, along with how they fitted in with the general development of the subject.

I decided to stop at four parts and 13 chapters. In the end what worries me is that there are only two women represented here. One of these has a minor role, but the work of one of them was central to Fermat's last theorem for a long period of its history. Unfortunately, she was very badly treated by the "boys" and the culture around her in that she was not given the recognition that was due to her. The only way I can think of absolving myself is to encourage someone to write a book about the 20th and 21st century women who have made great strides in mathematics and who are beginning to be treated better.

The Relay Race to Infinity

The book is purposefully called *The Relay Race to Infinity*. This is because I have written of the way that mathematicians take a part of mathematics and make a contribution to the topic, after which others take up the baton and run with it, and for others to do that again and again. This relay race that aims to cover more and more of mathematics will go as long as our species exists. This may not get as far as infinity but it is covering more and more each day.

Giants

I'll use the word "giant" from time to time possibly linking it to "shoulders". It would seem that the first time "shoulders" and "giants" were put together in the sense I use here, was around 500 BCE. But we have it in writing from 1159, when John of Salisbury wrote in his *Metalogicon*,

> Bernard of Chartres used to compare us to dwarfs perched on the shoulders of giants. He pointed out that we see more and farther than our predecessors, not because we have keener vision or greater height, but because we are lifted up and borne aloft on their gigantic stature.

For most people the source is probably Isaac Newton, who wrote in a letter to Robert Hooke (5 February 1675)

> If I have seen further, it is by standing on the shoulders [sic] of Giants

See Turnbull (1959), volume 1, p. 416

I hope this book helps you to see further.

A Note on Sources

This book draws on a lifetime's experience with mathematics from a wide variety of sources, some of which are hard to access today. To make life easier for today's reader my policy has been to cite things that are readily accessible to everyone. This means I use mainly online works. The three main sources I have used in all of the chapters are MacTutor

`https://mathshistory.st-andrews.ac.uk/`

for biographies, Wikipedia and online Britannica for mathematics. These sources are very informative and have enormous scope, but they should be used with caution. As (I hope) everyone knows, the internet is unreliable and impermanent. Wikipedia and Britannica articles may be revised at any time, so if, say, a MacTutor biography quotes a Britannica article, the quote may no longer exist.

For this reason, I have supplemented online references, where feasible, with references to real books. That does not necessarily mean you have to run to the library to check what I say, since many books are available online. Many of the classics are available at the Internet Archive:

`https://archive.org/`

and I will give occasional references to other books that are available online. But in any case, you like reading books, don't you? Like the one you are holding in your hands.

All the best,

The Authors

PART 1
Euclid

Chapter 1

Life and Times of Euclid

Fig. 1.1 Image of Euclid from Oxford statue, photographed by Mark A. Wilson, Wikimedia

1.1 Background

First it's important to say that the work of Euclid contains a great deal of mathematics and it's not possible to cover it all. This is especially because, as for all parts of this book, I have written a largish chapter here that gives some historical background to Euclid's time. So what have I tried to cover about Euclid's contribution to mathematics? This is in Chapters 2 and 3.

In brief, I have restricted myself to an outline of the *Elements* (first section of Chapter 2), followed by Euclid's geometry and what distinguishes it from other kinds of geometry. This includes discussing ruler and compass constructions, Pythagoras' theorem and its extensions, and the Platonic solids in the remainder of Chapter 2. Then in Chapter 3 we see Euclid's number theory, including the basic results on prime numbers and the Euclidean algorithm. Finally, I look at Euclid's legacy to the mathematical world.

In some places I prove a mathematical result, but I've tried to make these forays comprehensible to a wide audience. Nevertheless, if you get stuck anywhere, skip that piece. There should still be sufficient material without these parts for you to get a bird's eye view of events.

For this chapter it's fortunate that Euclid's *Elements* has been translated into English. As a first reference you might like to look at

`http://aleph0.clarku.edu/~djoyce/java/elements/elements.html`.

Then, if you want a specific Book of the *Elements*, click on its Roman numeral in the diagram at the top of the screen. You can then navigate, in turn, to a specific proposition. Any time I refer to a specific Book, or Proposition, it is by this method.

This work by David Joyce is a modernisation of the classic translation by Heath (1925), and it has additional notes that are valuable. But the problem I have had with it is that the language is still old-fashioned and it can make the maths difficult to understand. The additional notes can help though. A more modern translation by Richard Fitzpatrick, also beautifully illustrated and placed in parallel with the Greek text is at

`https://farside.ph.utexas.edu/books/Euclid/Elements.pdf`

This is a single file; that is, the whole *Elements* in both Greek and English, hence a little overwhelming. Still, I urge you to look at it and admire it.

Yet another admirable piece of work is by Nicholas Rougeux:

`https://www.c82.net/work/?id=372`

At this site you will find Rougeux's reproduction of the first six books of the *Elements*, made by Oliver Byrne in 1847, in which all proofs are basically sequences of coloured pictures, with a minimum number of words.

1.2 Introduction

Euclid and Alexandria

You have to worry about a man who, on the web, has his birthday listed as both "about 325 BCE" and "mid-4th century BCE". However, you have to realise that Euclid did live a long while ago and nobody then had birth certificates. So how do you get an idea of a birthdate at this distance in time? Clearly you need to look at every piece of extant evidence even though there is not a great deal of it. The best hint is in a 450 CE document that linked him with students of Plato. If you have some idea of how old Plato's pupils were, you could have put Euclid's age to within at least a reasonable boundary. I have to say though, that dating his birth to 325 BCE is a little optimistic, but "about 325 BCE" is probably as good as can be done until some never-before-seen evidence comes along. Assuming that Euclid was born in Alexandria, which was founded in 331 BCE, then 331 BCE gives us a lower bound for his birth year.

This is not the only problem around Euclid's origin: it's not even clear where he was born! The general opinion is Alexandria though. This is possibly based on the fact that the *Elements* and his other works can be reliably traced to that city.

Fig. 1.2 Map of the eastern Mediterranean from https://www.worldhistory.org/image/283/map-of-the-mediterranean-218-bce/, public domain

Now Alexandria is a Mediterranean port in Egypt (see Figures 1.2 and 1.3) whose site was chosen by Alexander the Great in 331 BCE. When he found the site, it was basically open land, but his choice had been considered very carefully because it was to be the site of his throne. Unfortunately for him, he never returned. Alexandria had two harbours. This was good for both war and peace. He also chose an architect Dinocrates, see

https://www.britannica.com/place/Alexandria-Egypt/City-layout

to plan and develop his capital.

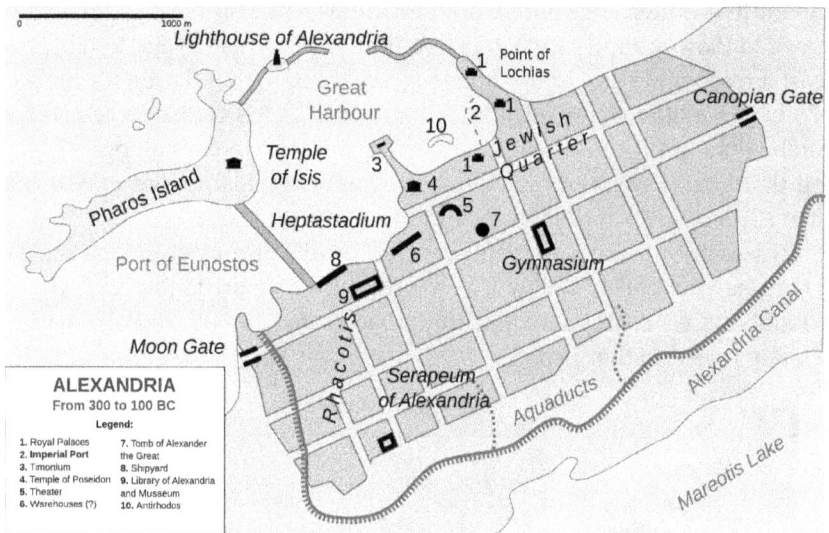

Fig. 1.3   Map of Alexandria at the time of the Ptolemies (Wikipedia, Mouseion)

The new city was carefully planned on a rectangular grid. Due care was made to make it pleasant with easy access to open areas. Special care was taken of the richer members of society. In order for them to have access to fresh water, tunnels were made that took water from the Nile directly to their homes. It is likely that Euclid shared this advantage.

Thanks to Dinocrates and to Ptolemy I and II, Alexandria quickly became a centre of commerce and a connection between Europe, Northern Africa and the Arab regions. Hence Alexandria thrived and developed as a culturally oriented city of Greeks, Egyptians, Jews and Arabs.

The city built a library that was enriched with the help of foreign ships. Alexandria took any books that were on board these ships and these were

copied and the copies were deposited in the library. Of course, the origi-
nals were returned to their owners. As a result, Alexandria had the best
library in the known world at that time. This library was part of the
Musaeum, a large institution that also included rooms where study and
discussions were held, as well as places for lectures. Scholars lived there
and they were given free board, meals and servants by the governing body.
Euclid was a tutor at the library. Because of its library and the *Musaeum*,
Alexandria was the place to which many ancient Greek intellectuals grav-
itated.

Alexandra was also famous for its Lighthouse, Figure 1.4, one of the
Seven Wonders of the Ancient World. Standing at 100 metres high, when
it was built around 280 BCE it was the largest building in the world. Three
earthquakes, the last in 1323 CE, finally destroyed it.

Fig. 1.4   The Lighthouse of Alexandria (Wikipedia, Lighthouse of Alexandria)

At this point there is something I should explain. In Euclid's time, the
name "Euclid" was common. For example, our Euclid lived at roughly
the same time as Euclid of Megara. (Megara is a place in Greece.) The
Euclid who wrote the *Elements* has often been confused with the man from
Megara. From now on you may assume that when I say "Euclid" I mean
the author of the *Elements*.

But because we have so few details about him personally, the question
that has to be asked is did Euclid ever exist and was there really just one
person who wrote the *Elements*? This is actually a question that has caused
much professional discussion and there are three proposed answers that
have been given some credence.

First, it could be that there wasn't a single person called Euclid who wrote the *Elements*. Maybe a group of people got together, wrote different sections of the book, and invented an author's name, "Euclid", that covered all of their work. An instance of this happened in the last century. After the First World War there was a shortage of senior academic mathematicians in France. There was also a shortage of books. So, a group of young French mathematicians got together and wrote some maths books and used "Bourbaki" as the name of the author (see Nicolas Bourbaki in Wikipedia). Having different writers contribute to such a project could explain the variation of the writing styles to be found in the *Elements*.

Second, Euclid might have had a group of followers and they all wrote pieces of the book and Euclid edited them or maybe just put his name to their work. This happened with Pythagoras and with famous artists in the Renaissance. It still happens today in various companies where the name of the company owes its origin to someone who did excellent work in a particular area, such as perfume. Again, writing done this way would be likely to vary in style, especially if the "author" was not a very good editor or perhaps was even dead.

The third possibility is the easiest to state and probably accept. Euclid was Euclid and he did exist about the time that the web suggests. This seems to be the best of the three suggestions based on current available evidence. What about the different styles? That can be put aside because people do change their style as they mature and as their emotions change. My book for example, changes depending on whether I'm talking about maths, or history or whatever, and there are two authors too. On the other hand, Euclid was taking other writers' mathematics and writing it in a logical progression. As a result, you might expect some variation in the styles of different parts of his work.

Oh, and if there is all this indecision about his birthday and his book, then you can hardly expect to find a portrait of him. Well, any portrait that accurately portrayed him. I apologise for the picture of him in Figure 1.1 which is certainly not accurate, but it's nice to have a face to hang the name to. On the other hand, if you don't like this face, you can find many more on the web, which are equally unreliable.

Having cast doubt on Euclid, there is no way to cast doubt on the book, the *Elements*. There are parts of original parchments extant as well as many copies and references to the work by other authors who lived when Euclid was supposed to live or in the years not too long after his death. I'll investigate this book and mention others by Euclid in a later section.

Having raised a controversy right at the start, I hope that I can stir up a few more so that you have an enjoyable time both in this and in later chapters.

So, long, long ago, there was a man and a book or two. The author was Euclid and the outstanding book of the "book or two" was the *Elements*. You may have heard about him/them (and this really is a plural "them"). The *Elements* was important because it

(1) put together all of the geometry, both two- and three-dimensional, known at that time;
(2) included number theory and irrational numbers; and
(3) used axioms and their logical consequences to establish results.

Euclid and a Conjectured Private Life

As Euclid was a tutor at the library it is likely that he had residence there for at least part of his life, but it is not clear whether he had an income or inherited wealth from his father.

It's worth looking at pictures and statues to get a good idea of what clothing was worn in Alexandria. For the local Greek community, their wear was influenced by customs in Greece. In ancient Greece, people wore an undergarment and a cloak. These were loose and free-flowing. Both sexes wore the same dress, but women wore their clothes down to the ankles while the men's went just to the knees. Clothing was largely made at home or nearby, with linen being used for summer and wool for the winter. The wealthier you were the more colour your dress was likely to be. Because of the difficulty in producing purple dye, purple clothes were only worn by the very rich. Clothes from overseas were also available for the richer section of society. It was acceptable for a man to be seen in public without any clothes, but womens' nudity was restricted to the public baths. See Wikipedia, Clothing in Ancient Greece.

Euclid undoubtedly wrote with a reed pen on papyrus. See

```
https://sites.dartmouth.edu/ancientbooks/2016/05/23/
the-writing-instrument-the-reed-and-quill-and-ink/
```

His ink would have been produced from burnt wood mixed in oil with gum from trees to help the ink stick together. In Egypt, iron oxide was added to give a red ink. Euclid might have made his pen and ink himself, but it would have been possible to buy them or get a slave to make them for him.

As for furniture, it was made of a variety of accessible materials including ivory, metals, reeds, stone, straw, and wood. Chests, stools, tables, and reed mats were common. For a look at furniture in the ancient Greek world, see Wikipedia, Ancient Furniture.

1.3 History

Wars

I prefer not to have any contact with wars, but it turns out that wars were going on at around the time of Euclid. I thought it might be interesting to look at these, and the people involved, and the consequences for them.

The **Warring States** were seven states of China (Figure 1.5) that fought to annex more land until they all united into the one country, China. Before the beginning of the period, 475 BCE, there was a great deal of infighting between various states and combinations of states with a view to increasing their size, strength, and status. This role of conquest and annexation was common in this period, not only in China, but in Europe (see the Punic wars) and across Persia and India (see Alexander the Great). The details of these Chinese wars can be found in Wikipedia, Warring States period.

Qin Shi Huang was King of Qin during this time and from the start of his reign the Terracotta Army began to be constructed. This was a gigantic effort to produce an army to guard Qin when he died and went to the afterlife. It contained terracotta images of soldiers from all ranks of the army. (See Wikipedia, Terracotta Army.)

Finally, in 221 BCE, Qin was the state that became victorious and Shi Huang became "The First Sovereign Emperor of Qin". Qin's reign was possible by the strength of his armed forces. But he made significant changes to the way that the country was run, many of which were designed to bring uniformity on straightforward matters to the country. For example, Qin standardised units of length, weight, currency and even axle widths! On a government level, feudalism was abolished and various administration units were established. To improve access across the whole of China, roads and canals were built. Further, the Chinese script used was made uniform across the country. Because of constant encroachment of tribes from the north and north-west, walls had been built for defense. These developed eventually into the Great Wall of China. However, it is likely that these were mostly propaganda to help the population feel safe and did not do a great deal to defend them from the north.

Fig. 1.5   The seven warring states (Wikipedia)

Qin Shi Huang was known to have banned and burned books as well as executing scholars. This certainly is true, but it is almost certain that this was exaggerated by the Han dynasty that followed.

This takes us back in time to Confucius. Can I say right at the start here that I have never really liked the way that British-speaking countries find it impossible to say the names of a foreign country properly, or at least try to do that. For example, surely we can all say "Paree" rather than "Paris"? And this habit appears to hold true for other countries. For example, Jesuit missionaries in the 16th century renamed the Chinese philosopher Kongzi, as "Confucius". Kongzi was a philosopher who died before the Warring States period. His philosophy was centred on "personal and governmental morality, correctness of social morality, justice, kindness and sincerity" (Wikipedia, Confucius).

Basically Kongzi taught his golden rule:

Do not unto others what you would not want done to yourself.

While Kongzi was not around when China was united, one of his followers, Meng Ke, 372–289 BCE was. (Of course, the Jesuit missionaries changed Meng Ke to Mencius.) Meng Ke developed Kongzi's philosophy further based on his belief that human nature is basically righteous and humane. (More on Meng Ke's philosophy can be found in Wikipedia, Mencius.) His version of Confucianism is now considered to be the standard one and this has made a lasting impression on East Asia to this day.

Alexander the Great

If you are the son of a king, then everyone knows your birth date down to a day, even if you were born in 356 BCE. The king was Phillip II, his domain was Macedon, his son was Alexander (see Figure 1.6), and he was born on the 20th or 21st of July in Pella, the capital of Macedon. That indecision of the day isn't the result of being unsure of the actual day. In fact, the actual day is not sure of itself. It is known precisely when Alexander was born. It was on the 6th day of Hekatombaion in the Attica calendar. This calendar is a lunisolar calendar which is based on both the Sun and the Moon. It turns out that it is difficult to translate the 6th of Hekatombaion to a precise date in our calendar.

Fig. 1.6 Alexander the Great in mosaic (Wikipedia, Alexander the Great)

Alexander wasn't called "Great" for nothing. His father was murdered in 336 BCE, which made Alexander King of Macedon. He had the ambition of owning a much larger kingdom than Macedon, and he immediately started a process that would eventually give him control of 5,200,000 km² of land that stretched from Macedon to Greece through the Middle East, around the Mediterranean to Egypt through all of Persia and as far as the Indus in India. He only stopped there because his troops had been fighting for 10 years and demanded to go home to their wives and families. On the other hand, the 80,000 horsemen, 200,000 footmen, 8,000 chariots and 6,000 elephants that were waiting on the other side of the Ganges, plus the width and depth of the river itself, may also have had something to do with Alexander's decision to stop there and go home.

Incidentally, the area under Alexander's control at this point (outlined in green in Figure 1.7) would have made his empire the seventh largest by area in the world today, after Russia, Canada, China, the USA, Brazil and Australia. Not bad for something achieved in 10 years.

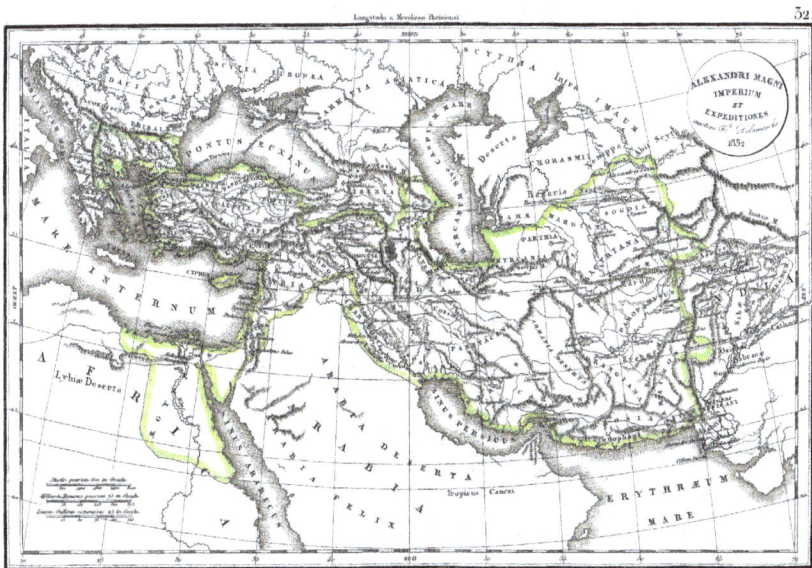

Fig. 1.7   Alexander's empire (Wikimedia, 1832 Delamarche Map of the Empire of Alexander the Great)

Alexander might have been called "Great" solely because of his conquests, but these weren't achieved solely by luck or by the number of soldiers he had under his control. There were times when his numbers on the field were inferior to his enemy. His conquests owed a great deal to his strategic use of both locality and arms. As locality goes, he might approach the enemy via an unexpected route, similar to Wolfe's outflanking of the French at Quebec. Regarding arms, he developed an idea of his father's called the Macedonian phalanx (see Figure 1.8). This involved using long spears by men in a square formation (Wikipedia, Macedonian phalanx). In the Battle of Gaugamela, he used both of these ideas (see Wikipedia, Battle of Gaugamela). As the result of his skill in warfare, he fought 20 national foes over 10 years and never lost (see battle record in Wikipedia, Alexander the Great).

Fig. 1.8   A Macedonian phalanx (from Wikipedia, Macedonian phalanx)

In 323 BCE Alexandra died, possibly as the result of poisoning, though it has been suggested that his death was caused by excessive drinking over a period and/or wounds incurred in battle. If it was poison, it is not clear whether the poison was deliberately administered or whether it was from drinking polluted water from the local river. Whatever it was, he had no heirs until a son was born to his wife after he died. There was some attempt to have this child brought up as King of Macedonia, but that failed. During his lifetime, Alexander had given different regions to his best generals to administer. Now they took over these areas and ruled them in their own names. This led to 40 years of war and finally the empire was divided into four regions, including the part of Egypt ruled by Ptolemy (his line ending with the death of Cleopatra in 30 BCE). Alexandria was the capital of this pharaohship and it quickly grew a strong population from Greece. This included many academics, such as Euclid and Archimedes.

The **Punic Wars** occurred between 264 BCE and 146 BCE and were between Carthage and the Roman Empire. Figure 1.9 shows how things stood at the beginning.

Fig. 1.9　The area involved in the Punic Wars (Wikipedia, Punic wars)

Carthage was actually a city in what is now Tunisia. Before the wars, Carthage had developed into a major Mediterranean power and so had become a threat to the Roman Empire. The war was partly motivated by Rome's desire to extend its domain by occupying Sicily. The island belonged to Carthage at the time and it wasn't very happy with Rome's attitude. Thus began the Punic Wars that were fought across the Mediterranean and largely engaged the naval power of both combatants. The wars ended with Rome not only winning Sicily but gaining land in Spain and North Africa and driving Carthage back to its origin in Tunisia. There is a valuable animation of the changing of the land in Wikipedia, Punic Wars, which shows the development of the conflict.

Incidentally, it is now only about an 8-hour drive from Carthage to Béjaïa, Algeria, where Fibonacci (Part 3) lived over a millennium later.

1.4   Philosophers and Mathematicians

Plato

Plato was born in 427 BCE and died in 347 BCE. But there is some debate about these dates, with the first variously given as 428/427 or 424/423 and the second as 348/347. Even though Plato was from a wealthy family it seems still difficult to locate his birth and death years with any accuracy. However, what really matters is his achievements.

He was born and died in Athens but travelled extensively for a variety of reasons. One of these was as a soldier and another was to get away from Athens "democratic" leaders and "see the world" of Egypt, Cyrene, and Italy, including Sicily. On these travels Plato picked up the mathematics attributed to Pythagoras and his followers. (By this time little trace of Pythagoras remained, but mathematical folklore gave him credit for results such as Pythagoras' theorem.) Although Plato himself never became a mathematician of great standing, his political and philosophical interests led him to see the importance of the subject. From his contact with Pythagoras' work, Plato determined

> ... that the reality which scientific thought is seeking must be expressible in mathematical terms, mathematics being the most precise and definite kind of thinking of which we are capable.
>
> G C Field, *The Philosophy of Plato* (Oxford, 1956), p. 34

This realisation guided the evolution of science from that day on.

When Plato returned to Athens after touring parts of the Mediterranean, he established an Academy which was to be a place for research and teaching in philosophy and science. But mathematics was fundamental to his thinking as well as being considered to be part of science. Here is his program for the Academy's curriculum.

> ... the exact sciences – arithmetic, plane and solid geometry, astronomy, and harmonics – would first be studied for ten years to familiarise the mind with relations that can only be apprehended by thought. Five years would then be given to the still severer study of "dialectic". Dialectic is the art of conversation, of question and answer; and according to Plato, dialectical skill is the ability to pose and answer questions about the essences of things. The dialectician replaces hypotheses with secure knowledge, and his aim is to ground all science, all knowledge, on some "unhypothetical first principle".
>
> MacTutor biography of Plato

On the research side of the Academy programme, Plato concentrated on the idea of proof and insisted on accurate definitions and clear hypotheses. This laid the foundations for Euclid's systematic approach to mathematics in the *Elements*.

It's interesting how words enter the language. The word "Academy" comes from Akademos, a hero of Greek mythology. Plato situated his academy in a garden dedicated to Akademos; see, for example Netz (2022), p. 392. So in fact Plato's Academy was not a grand building nor a building of any sort, just a garden. This means that when you read that over the "door" of the Academy there was the sign:

Let no one unversed in geometry enter here,

there might just be an error somewhere. Indeed, there is some doubt as to whether there was ever such a sign though possibly it was on a tree somewhere. It's worth looking into this. In fact, different references I've looked at seem to have slightly different ideas on the subject.

Plato was strongly influenced by his teacher Socrates, who was put on trial and forced to commit suicide, apparently in Plato's presence. It was on Socrates that Plato based his writings called Dialogues. The name comes from the fact that Plato set them up as discussions between Socrates and others. In the Dialogues, Socrates is thought to be expressing views that were Plato's. Among the topics covered in these discussions are religion and science, reality, nature, human nature, and love and sexuality.

Plato's Academy existed until 529 CE, well after Plato's death! It was finally closed by the Christian Roman Emperor Justinian because he claimed that it was heathen.

Theaetetus

Apparently Theaetetus had bulging eyes and a snub nose. I only say this because he shared these attributes with Socrates, his teacher. The eyes and nose made him less handsome than he might otherwise have been. Theaetetus was also a friend of Plato who wrote two Socratic dialogues starring him. From this we know that Theaetetus had a "beautiful mind" and was a "perfect gentleman".

Theaetetus was born in Athens in about 417 BCE. He holds his place here because of the part he played in the *Elements*. Book X, on irrational numbers was essentially worked out by Theaetetus and it is considered to be the deepest part of Euclid's book. It covers irrational numbers that arise in ruler and compass constructions, up to the complexity of square roots inside square roots. Book X is admired for the thoroughness and depth of its proofs. Theaetetus's work also contributed to the Platonic solids treated in Book XIII. Though the tetrahedron, cube, and octahedron were studied by the Pythagoreans, the dodecahedron and icosahedron were attributed to Theaetetus. They involve lengths described by square roots inside square roots, such as $\sqrt{(5 + \sqrt{5})/2}$. Further, he was able to show how all of these solids could be constructed and inscribed in the sphere.

Unfortunately, Theaetetus took part in the Athenian war against Corinth in 369 BCE. He seems to have fought well, but he was injured and as a result of his wounds he contracted dysentery and died.

For more information I recommend reading Theaetetus, MacTutor.

Aristotle

Aristotle was born about 384 BCE in Stagirus, Macedonia, Greece and died around 322 BCE in Chalcis, Euboea, Greece. As a teenager, Aristotle joined Plato's Academy and remained there for 20 years. In this time, he evolved from a student to a teacher. His teaching covered a very large range of topics. It is believed by some authorities that at one point he taught the young Alexander the Great.

Aristotle hoped to be the head of the Academy, but he never achieved this goal. He left Athens for a period in which he did unprecedented work on zoology and marine biology, producing a large collection of specimens.

During the period of Alexander's conquests, Aristotle moved back to Athens where he set up his own Academy called the Lyceum. Unlike the Academy, the Lyceum was an open institution where anyone could come and listen to lectures.

Logic, as a formal discipline, was invented and developed by Aristotle. He formalised the notion of **syllogism**. This is a means of deduction that starts with two propositions, or statements, that together imply a result. Below we give two propositions that imply a conclusion. Overall, the three propositions form a syllogism.

> Every Greek is human.
> Every human is mortal.
> Therefore, every Greek is mortal.

The "every Greek" of the first proposition is a subset of "every human" in the second proposition, so the third line must be true.

But there is more to logic than syllogisms and Aristotle not only recognised this but considered other logical situations. In fact, he developed logic to an extent that in some ways was not surpassed until the 19th century.

Aristotle made some progress with the infinite by distinguishing continuous lines, for example, by the fact that between any two points on such lines there were always further points. But by the same reasoning time and motion are also continuous. Between any two times there are other times. Between any part of a motion there are other points of that motion.

But Aristotle also continued to break up a continuous line. A continuous line can be divided up into continuous lines, which again can be divided up. And this process can be continued indefinitely. So, the line is infinitely divisible. No matter how often a continuous line is divided, it can be divided even further. This is the idea of **potential infinity** – I can always take one step more. For instance, think of the whole numbers. They start 1, 2, 3, and then there is a next one. What is the largest whole number? Well suppose I thought it was 2809. Then clearly 2810 is bigger. Fine then, perhaps 22810 is the biggest number. But again I can add 1 to get 22810 + 1, which must now be the biggest number! Obviously this process goes on forever. So the whole numbers are certainly potentially infinite. However, Aristotle didn't believe in an **actual infinity** – one that existed as an entity. As he writes in his *Physics* (see for example MacTutor):

> But my argument does not anyhow rob mathematicians of their study, although it denies the existence of the infinite in the sense of actual existence as something increased to such an extent that it cannot be gone through; for, as it is, they do not need the infinite or use it, but only require that the finite straight line shall be as long as they please. ... Hence it will make no difference to them for the purpose of proofs.

The idea that an actual infinity could not exist was held by most mathematicians right up to the 19th century and they certainly managed to get along without it. Basically, as Aristotle said, mathematicians didn't need it – that is, until they did, as Cantor discovered. You may like to look at Wikipedia, Actual infinity, too. But there are still mathematicians who choose to do mathematics without it. These are the intuitionists. (Wikipedia, Actual infinity also discusses them.)

Overall, Aristotle made significant contributions to many areas of knowledge. In Britannica (`https://www.britannica.com/biography/Aristotle`) it is said of Aristotle that:

He was the author of a philosophical and scientific system that became the framework and vehicle for both Christian Scholasticism and medieval Islamic philosophy. Even after the intellectual revolutions of the Renaissance, the Reformation, and the Enlightenment, Aristotelian concepts remained embedded in Western thinking.

Aristotle was famous too for his model of the Universe (Figure 1.10). This was held in high regard for 2,000 years or so after his death. For a start, Aristotle gave a clever argument that the Earth is round (don't let anyone tell you Christopher Columbus was the first to think of this!). He realised that the Moon passes through the Earth's shadow whenever an eclipse of the moon occurs, and he noticed that the shadow of the Earth on the Moon is *always a part of a circle*. Only a round object always has a round shadow, so the Earth must be round. Other signs of the Earth's

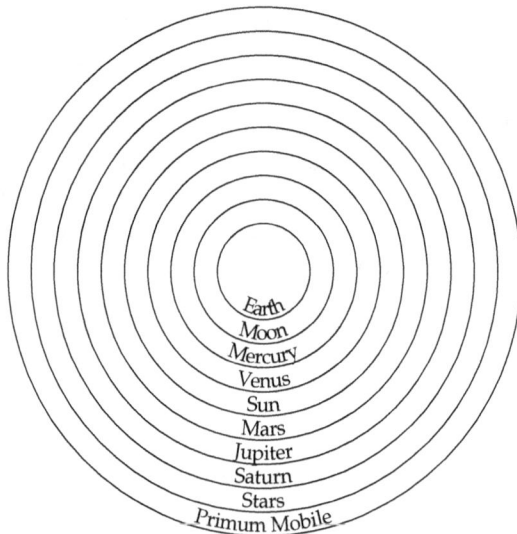

Fig. 1.10   Aristotle's model of the universe

roundness were also noticed in ancient times: for example the way ships disappear over the horizon, and the fact that different stars are observed at more southerly locations.

But it was not until astronomers began to know more of the heavens, that they became suspicious of the rest of Aristotle's model, which had the other bodies circling around the Earth. It was through work by people such as Copernicus, and finally Galileo, that Aristotle's geocentric view of the universe, became the heliocentric view of the solar system that we know today.

In Figure 1.10, I have shown a copy of Aristotle's model. It looks very strange compared with what we currently believe. First, all of the heavenly bodies moved in circular orbits around the Earth. Second, the Sun has no special position. The Moon and the Sun were gathered together with the planets in the inner regions of Aristotle's Universe. Outside of these are "fixed stars" and the "Sphere of the Prime Mover". Indeed, he thought that all stars were fixed, otherwise there would be some evidence of their relative movement on Earth. But who or what was the Prime Mover and what was the Primum Mobile sphere that the Prime Mover moved?

Eratosthenes

Eratosthenes (roughly pronounced Error-toss-tha-knees) invented a mathematical sieve and worked on things we would call geography today. In geography his most spectacular achievement was a determination of the Earth's circumference, accurate to within a couple of percentage points.

He was born in 276 BCE at a place called Cyrene (now called Shahhat), which is presently in Libya. However, he wasn't a Libyan but a Greek! This is a little surprising as mainland Greece is on the other side of the Mediterranean. In fact, Greece founded a country of five cities, well before Eratosthenes was born, with Cyrene as the capital. The Greek influence of the country and other regions in the north of Africa was strengthened when they were conquered by Alexander the Great in 332 BCE. So Cyrene was still under Greek control when Eratosthenes was born. Thus his early childhood was spent in a cultivated Greek city, and he was brought up in much the same way as he would have been in Greece itself. At later stages he furthered his education in both Athens and Alexandria.

Now I should say what *Eratosthenes' sieve* is and what it does. First look at the listing of the numbers up to 100 in Figure 1.11. What I am about to do is to find all of the prime numbers from 1 to 100. And, like Eratosthenes, I'm going to do it systematically.

- Now 1 is not considered to be a prime, so I'll cross it out.
- But 2 is a prime, in fact it is the only even prime, so I won't cross it out. However, I'll now cross out all other multiples of 2 because they can't be primes. Thus I cross out all but one of the even numbers.
- Now there is a number next to 2 that hasn't been crossed out. So that

means that 3 is a prime. From here I'll cross out all other multiples of
3. Some of these, like 6, have already been crossed.

- Just keep going. Having done my worst with 3, I look for the next
  number that is not crossed out. Since neither 2 nor 3 divide 5, then 5
  is our next prime. I won't interfere with 5 itself, but I will cross out all
  other multiples of 5. I'll leave you to do that.

- And I'll also leave you to find the next prime and delete all its multiples. Keep going until the next prime you find has all multiples less
  than 100 already crossed out. You have then found all of the primes
  up to 100. They are the numbers not crossed out in Figure 1.11.

$$
\begin{array}{cccccccccc}
\not1 & 2 & 3 & \not4 & 5 & \not6 & 7 & \not8 & \not9 & \not{10} \\
11 & \not{12} & 13 & \not{14} & \not{15} & \not{16} & 17 & \not{18} & 19 & \not{20} \\
\not{21} & \not{22} & 23 & \not{24} & \not{25} & \not{26} & \not{27} & \not{28} & 29 & \not{30} \\
31 & \not{32} & \not{33} & \not{34} & \not{35} & \not{36} & 37 & \not{38} & \not{39} & \not{40} \\
41 & \not{42} & 43 & \not{44} & \not{45} & \not{46} & 47 & \not{48} & \not{49} & \not{50} \\
\not{51} & \not{52} & 53 & \not{54} & \not{55} & \not{56} & \not{57} & \not{58} & 59 & \not{60} \\
61 & \not{62} & \not{63} & \not{64} & \not{65} & \not{66} & 67 & \not{68} & \not{69} & \not{70} \\
71 & \not{72} & 73 & \not{74} & \not{75} & \not{76} & \not{77} & \not{78} & 79 & \not{80} \\
\not{81} & \not{82} & 83 & \not{84} & \not{85} & \not{86} & \not{87} & \not{88} & 89 & \not{90} \\
\not{91} & \not{92} & \not{93} & \not{94} & \not{95} & \not{96} & 97 & \not{98} & \not{99} & \not{100}
\end{array}
$$

Fig. 1.11   Eratosthenes' sieve

Once you have done that, I suggest that you look at Wikipedia, Sieve
of Eratosthenes. There you'll find an animation that goes through all of
the steps necessary to find the prime numbers less than 100.

Just to see where Eratosthenes fits historically, here are a few people
who live chronologically around him. For example, Alexander the Great
(Alexander), preceded Eratosthenes by roughly 80 years. Archimedes,
who is supposed to have run through the streets shouting "Eureka", overlapped with Eratosthenes. The Qin Dynasty of China also existed during
Eratosthenes' life. On the other hand, Julius Caesar (Wikipedia, Caesar),
postdated our subject by about 100 years.

1.5   Other Books by Euclid

Euclid wrote several other books apart from the *Elements*. Five of these books still exist in some form, but many have been lost. We know about the lost books because they are referred to by other authors. Generally, all Euclid's books are written in the same style as the *Elements*, using definitions and axioms from which follow theorems. I list these books with some comments.

Catoptrics. This book is about a mathematical theory of plane and spherical concave mirrors. However, it is likely that this book was actually written by Theon of Alexandria (Wikipedia, Theon of Alexandria). Theon lived in Alexandria, hence the name, roughly during the fourth century CE. It is interesting that Hypatia (Wikipedia, Hypatia), Theon's daughter, is one of the very early female mathematicians. It is not clear that Hypatia made any significant original contribution to mathematics, but she did make contributions to its development by her work on commentaries on the classic works of the time.

Data. This is not of great significance as it appears to be very strongly related to the *Elements*, Books I to IV.

On Division of Figures. It is about dividing geometric figures into parts; for example, dividing a triangle into two equal parts by a line parallel to one of its sides. Fibonacci, whom we will meet in Chapter 7, also wrote a book on such problems.

Optics. Here Euclid made contributions to perspective and the theory of vision. A number of Renaissance artists worked with linear perspective because of their access to this book.

Phaenomena. The topic of this book was spherical astronomy. Euclid's life overlapped with that of Autolycus of Pitane who wrote a book that is very similar to this one.

Others. Several books of Euclid have been lost. One of them, *Conics*, has connections to that of Apollonius of Perga. His book was called *Conics* too. It is possible that Euclid's *Conics* is a first draft of the early chapters of Apollonius' one. Later Apollonius may have rewritten or added to Euclid's book and then added more chapters on the topic. It is possible that for this reason, Apollonius' book survived while Euclid's just faded away.

1.6 Euclid's Legacy

In the *Elements*, Euclid collected together results and proofs from ancient Greek mathematicians. The book also includes proofs by Euclid of their material (see, for example, his proof of Pythagoras' theorem and the proof that there are precisely five Platonic solids). The *Elements* contains two- and three-dimensional geometry as well as number theory. Some of the new material in the book is the Euclidean algorithm that finds the greatest common divisor of two whole numbers and the first proof that there are an infinite number of prime numbers.

Euclid's legacy is based on the *Elements*. There are two main aspects that have been valuable for mathematics as a whole. First, the thing that really stands out is the way that he logically develops theorems starting from definitions and axioms (and proved theorems). This way he has provided a standard format for the current method of presenting results for research papers, as well as modern mathematical books on specific topics.

Second, the book is presented clearly and is considered a master text-book. When I was a student in the mid-20th century, the geometry I learnt was still being taken directly from the *Elements*. (The old-fashioned translations don't make it as readable today, though, as it might otherwise be.)

Chapter 2

Geometry

2.1 An Outline of the *Elements*

The *Elements* included original work of Euclid, especially

(1) A proof of Pythagoras' theorem.
(2) The infinitude of primes and the fundamental theorem of arithmetic.
(3) The beginning of what is still the method of writing proofs.

But certainly there are parts that are the work of others, such as Pythagoras and Theaetetus.

General information. The aim of the *Elements* is to develop sections of mathematics as they were known in Euclid's time and to set out the work systematically and logically with axioms and proofs of theorems. The *Elements* flow so that each of the Books proceeds from axioms to theorems from which further results can be obtained. For example, Book I starts with the equilateral triangle and ends with Pythagoras' theorem.

I should like to say here that there are two things that made it difficult for me to read the *Elements*. First there was the use of language that is not used in current mathematics. It takes a while to get used to this. And there is the "language" of algebra. Of course, today, algebra is written with unknowns and not with pictures. It could be said that Euclid was doing algebra but used geometrical objects as his variables. But it's not surprising that I found Euclid's method harder to follow – I'm just not used to it, and it is far more longwinded than algebra today.

But the English version of the *Elements*, already mentioned in Section 1.1, is of immense help:

`http://aleph0.clarku.edu/~djoyce/elements/elements.html`

I suggest that you browse through at least some of the parts of this version, which I discuss below, in order to get a feeling for Euclid's book. One thing that you might find useful is the Guide given for each of the propositions. This gives a commentary on the proof and talks about what Euclid is trying to achieve.

There is no doubt that the *Elements* is an important and influential book in the history of mathematics – probably *the* most influential. In the way it develops mathematics, it has influenced the subject to the present day. As a teaching book it is incredible – the main textbook in maths up to at least the late 19th century. Looking back on it now, it is clear that the geometry curriculum in my secondary schooling still owed much to Euclid. But no one seems to appreciate the value of geometry in school anymore.

The *Elements* is made up of 13 books. I have laid out the contents in the list below.

Books I to VI: two-dimensional geometry.

Book I & II: basic properties of triangles, quadrilaterals, parallel lines.
Book III: properties of circles.
Book IV: essentially the work of Pythagoras and his followers.
Book V: involves Eudoxus' material on proportion.
Book VI: gives applications of Book V to two-dimensional geometry.

Book VII to IX: number theory.

Book VII: introduction to number theory and the Euclidean algorithm.
Book VIII: geometric series.
Book IX: number theory.

Book X: irrational numbers based on Theaetetus.

Books XI to XIII: three-dimensional geometry.

Book XI: basic definitions for the next books.
Book XII: following Eudoxus: with regard to radius, circles are measured by squares and spheres by cubes.
Book XIII: properties of the regular polyhedra (the Platonic solids), again due to Theaetetus.

2.2   Ingredients of the *Elements*

I'm not a big fan of cooking metaphors, but the *Elements* is made from certain, well, ingredients that must be laid out and understood at the start.

Definitions, postulates, common notions, and propositions. These words occur in most of the Books of the *Elements*, so it is important to say something about them first. In general, the *Elements* used a logical development from definitions, postulates and common notions to propositions, each following from earlier propositions or postulates. This approach was certainly influenced by Aristotle.

A **definition** should be precise. It uses a word to describe an object or a concept of note to avoid repeating what the object or concept is every time it is referred to. For example, you wouldn't want to say "look out there is a four-wheeled object that carries passengers and is driven by an engine of some sort that is about to run . . . Oh well." While a quick "car!!" might have saved someone.

What do you think these phrases define?

> that which has no part;
> a breadthless length;
> a line which lies evenly with the points on itself;
> a surface which lies evenly with the straight lines on itself;
> the inclination to one another of two lines in a plane;

In order, they define **point**, **line**, **straight line**, **plane**, and **plane angle**. It's probably hard to understand some of these. How would you define them? Don't bother! I think you will agree that points and straight lines *cannot* be defined in terms of anything simpler, so they should be left *undefined*. After all, definitions have to start somewhere, and it happens that all the other objects Euclid wishes to discuss can be defined in terms of points, lines, angles, and planes. For example, a **right angle** is one of the angles made "when a straight line standing on a straight line makes the adjacent angles equal to one another".

A **postulate** is a statement that is assumed to be true. This is also known as an **axiom**. As with definitions, we have to start somewhere in proving statements, and Euclid's postulates are simple "self-evident" statements about points, lines, and other basic notions, from which he intends to derive all other true geometric statements. For example, Postulate 1 of Book I is

> To draw a straight line from any point to any point.

You can see here one of the peculiarities of Euclid's language. Instead of saying a straight line *exists* between any points, Euclid says you can *draw* it. From the beginning, the *Elements* is not only about what exists, but what can be drawn, and he assumes that the drawing instruments are the **ruler** (for drawing line segments) and the **compass** (for drawing circles).

Postulate 2 says that any line segment can be "extended", which is a cautious way of saying that lines are infinite.

Postulate 3 says that a circle can be drawn with any given centre and radius.

Postulate 4 says something that may strike you as strange:

> That all right angles equal one another.

You want to say "of course, because all right angles are 90°" but hold your horses! In the definitions, degrees have not been defined, only right angles. The fact that right angles can be defined and are equal makes the right angle a natural unit of angle measure, and indeed Euclid measures all angles as multiples of the right angle. For example, we will see later that the angle sum of a triangle is two right angles.

Euclid does not use numbers to measure lengths, areas, volumes or any other geometric quantity. (Indeed, there is *no* natural unit of length, unlike the natural unit of angle, the right angle.) He shows only that certain quantities are *equal*, using some "common notions" about equality that we will see shortly. There is a reason for avoiding numbers – **irrationality** – which I will say more about later, but in any case it is remarkable that geometry without numbers is even possible.

Postulate 5, known as the **parallel postulate**, is the most interesting axiom and the one that has caused the most investigation. Here is one translation of Euclid's statement:

> If a line segment intersects two straight lines forming two interior angles on the same side that are less than two right angles, then the two lines, if extended indefinitely, meet on that side on which the angles sum to less than two right angles. (Wikipedia, Parallel postulate)

This postulate is also being as cautious as possible, not assuming that lines are infinite, but merely allowing them to be extended until they meet. The postulate says that the lines $\mathcal{L}$ and $\mathcal{M}$ *will* meet if the line $\mathcal{N}$ crossing them in Figure 2.1 makes the angles α and β such that $\alpha + \beta < 180°$.

The term "parallel postulate" is also applied to a different statement: *that, given a line and a point not on that line, there is a line through the point not meeting the first line (when extended indefinitely).* This statement, called **Playfair's axiom** after the Scottish mathematician John Playfair, is equivalent to the one involving angles but obviously simpler and more economical

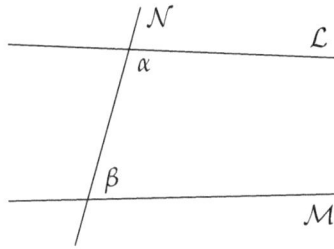

Fig. 2.1   The parallel postulate

in concepts. However, Euclid's version of the axiom is more immediately fruitful: for example, it leads to an easy proof that the angle sum of any triangle equals two right angles.

Still, no matter how you say it, the parallel axiom is awkward, and many mathematicians hoped to avoid it by showing that it follows from the other, more straightforward, axioms. All such attempts failed, and eventually it was shown that the parallel axiom does *not* follow from Euclid's other axioms, because there are geometries in which the first four postulates hold but the parallel axiom does not. We say more about this in Section 2.8 on **non-Euclidean geometry**.

But do you see how things in maths grow? This growth may take centuries, but in the development of geometry we have come from good old Euclidean geometry to all sorts of non-Euclidean geometries, some of which turn out to describe the universe we live in. Often someone with a baton nudges maths along just for maths sake. They play with something that interests them and they come up with a pretty result. Maths has moved forward, but maybe there is no benefit for the person in the street. Then later, another person fits the maths to the people's needs.

Sorry, I was going to tell you about common notions. I know you have been waiting patiently while I have wandered into all this modern geometry stuff and raved on about how maths goes forward.

Common notions are assumptions that Euclid makes in order to prove certain things are equal. They are really axioms about equality, addition, and subtraction. For example, common notions 1, 2, and 3 in Book I are

Things which equal the same thing also equal one another.
When equals are added to equals, the wholes are equal.
When equals are subtracted from equals, the remainders are equal.

Note that common notions are not restricted to any particular part of mathematics; they are statements about equality and very broad notions of "adding" and "subtracting". You might say that they are statements about equations, though Euclid did not really think in those terms.

Finally, **propositions** are statements to be proved, or **theorems**. They are proved using the definitions, postulates and common notions. I discuss Euclid's most basic propositions in the next section. You may also want to look at the Guide to each proposition in the Clark University *Elements*, which has notes pointing out where definitions, postulates and common notions are used in its proof.

2.3   Triangles and Other Constructions

If you haven't used rulers and compasses together for some time, it might be useful to play around with them and see what you can do. Draw some straight lines, both long and short, and make sure that some intersect. Try to draw a set of parallel lines. Make some circles of varying sizes and allow some circles to intersect. Can you make a face of a person with a circle and some parts of circles?

It would be a good idea to follow the construction steps where they occur in propositions, below or in the Clark University *Elements*. Satisfy yourself that you can produce the construction and that you understand why it works.

Proposition 1: To construct an equilateral triangle – that is, one with three equal sides – on a given finite straight line.

In a modern book, this might be written as:

Theorem 1: Let AB be a line segment. Then there is an equilateral triangle with AB as one of its sides.

Proof: By Postulate 3, we can draw a circle with centre A and radius equal to AB, and a circle with centre B and radius AB (Figure 2.2). These circles intersect at C and D, respectively. Draw the lines AC and CB.

Now $AB = AC$ since they are both radii of the circle with centre A. And $AB = CB$ since they are both radii of the circle with centre B. So $AC = AB = CB$, since "things which equal the same thing equal each other". Hence triangle ABC is an equilateral triangle. Q.E.F.

The letters Q.E.F. here stand for *quod erat faciendum* which is Latin for "which was to be done". This always signifies the end of a proof of a

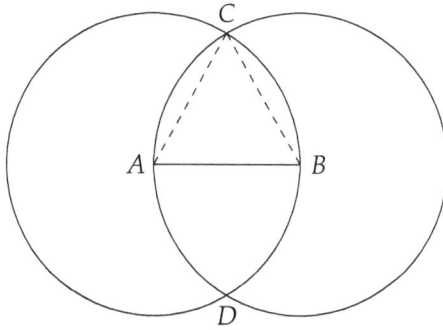

Fig. 2.2 Construction of an equilateral triangle

construction. On the other hand, if you see Q.E.D., it abbreviates the Latin for "which was to be demonstrated" which is *quod erat demonstrandum.* Again, it identifies the end of a proof.

One thing to note about the construction for Theorem 1 is that it also proves that ABD is an equilateral triangle. The proof illustrates the role of the compass in Euclid's geometry: it is a machine for making equal line segments. And to *prove* that certain line segments are equal we appeal to the common notions. What about proving equality of other objects, such as angles or triangles? Euclid's first theorem about equality of triangles is his Proposition 4:

> If two triangles have two sides equal to two sides respectively, and have the angles contained by the equal straight lines equal, then they also have the base equal to the base, the triangle equals the triangle, and the remaining angles equal the remaining angles respectively, namely those opposite the equal sides.

Figure 2.3 may help to explain what is going on.

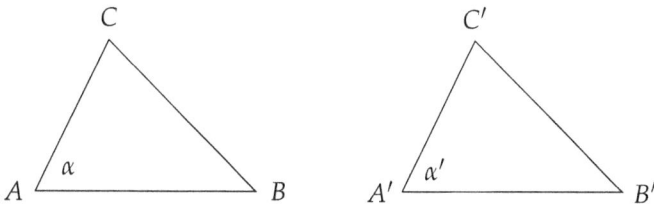

Fig. 2.3 The side-angle-side property of triangles

Euclid is claiming that if side AB = side $A'B'$, angle α = angle α', and side AC = side $A'C'$, then the triangles ABC and $A'B'C'$ are equal in all respects. They are what we now call **congruent**.

I don't call this proposition a theorem because, alas, Euclid's "proof" does not follow from his axioms. He smuggles in the notion of "moving" one triangle on top of another, but his axioms say nothing at all about "moving". In fact, Proposition 4 should really *be* an axiom itself, because otherwise we have no way to prove triangles congruent. You may know it as the property of triangles called SAS ("side, angle, side"), often used in high school geometry to prove congruence by showing that two triangles agree in two sides and the included angle.

SAS also gives a way to prove equality of angles, at least when the angles occur in triangles. The classic example is Euclid's Proposition 5:

> In a triangle with two equal sides (called an **isosceles** triangle), the angles opposite to the equal sides are equal.

Euclid's proof of Proposition 5 is fine, so I will call it a theorem. However, I will take the liberty of giving another proof, due to a later Greek mathematician called Pappus. The Pappus proof is so short you will miss it if you blink.

Theorem 2. If $AC = BC$ in triangle ABC, then the angle at A equals the angle at B.

Proof. Consider the two (!) triangles CAB and CBA in Figure 2.4, which of course are just two views of the triangle ABC. By assumption,

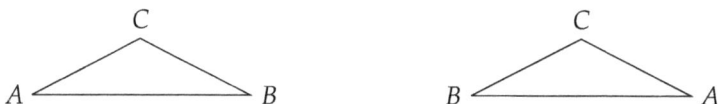

Fig. 2.4 Two triangles

left side CA of triangle CAB = left side CB of triangle CBA,
right side CB of triangle CAB = right side CA of triangle CBA.

Also, the angle included by the left and right sides is the same in both triangles; namely, the angle at C.

Therefore, by SAS, the corresponding angles in the two triangles are equal. This means the angles at A and B are equal.   Q.E.D.

Got that? Pappus is applying SAS to two triangles which are *both* the triangle we started with, but viewed in two different ways! (In the 1950s there was a flurry of excitement when a "geometry theorem-proving machine" produced this proof. It was at first thought to be original – until someone found it in the work of Pappus from nearly 2,000 years ago.)

In his Proposition 8, Euclid gives a dodgy "proof", similar to his proof of Proposition 4, of the principle known as SSS ("side, side, side"). This says that if two triangles have corresponding sides equal, then corresponding angles are equal too. He needs SSS immediately to prove his next Proposition, showing that any angle can be bisected. And right after that he uses SAS to show that a line segment can be bisected. Both bisections are quite easy to do with ruler and compass (try it!), but *proving* that the constructions are correct is entangled with the congruence principles SAS and SSS, which even Euclid did not completely understand.

Figure 2.5 shows how Euclid bisects an angle at the point D. He does something mathematicians love to do: "reduce the problem to one solved previously". In this case, Euclid uses his previous construction of the equilateral triangle, so we need not repeat how that was done.

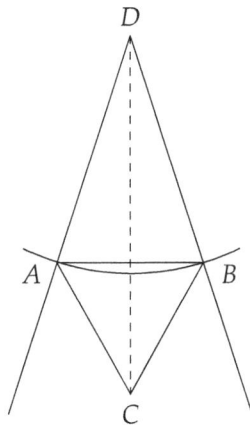

Fig. 2.5   Bisecting an angle

The first step creates points A and B on the lines through D such that $DA = DB$. This is done by drawing a circle with centre D, meeting the lines at A and B. Then construct the equilateral triangle ABC as in Proposition 1. Finally, connect C to D, which divides the angle at D into two.

To prove that the two angles just created at D are equal, apply SSS to show that the triangles DAC and DBC are congruent. The corresponding sides DA and DB are equal by the first step; the corresponding sides AC and AB are equal by the equilateral triangle construction; and the side CD is common to both triangles. Therefore, by SSS, the two angles at D are equal.

2.4   Angles and Parallels

So far we have not used the parallel axiom. We will certainly need it to make objects with parallel sides, such as squares and parallelograms, whose very existence depends on the parallel axiom. But we also need it to find the angle sum of a triangle. This theorem is Proposition 32 of Book I in the *Elements*, and it depends in turn on two more obvious facts, which follow easily from the parallel axiom and the definition of right angle.

- If a line across two parallel lines makes the angles α and β shown in Figure 2.6, then $\alpha + \beta =$ two right angles.

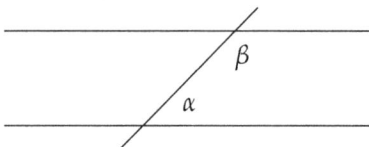

Fig. 2.6   Angles that sum to two right angles

- In the situation just described, angles α and β also appear at the positions shown in Figure 2.7. The two equal angles marked α are called **alternate** angles, as are the two angles β.

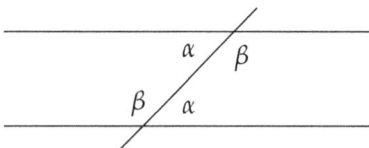

Fig. 2.7   Alternate angles are equal

Theorem 3. If α, β, γ are the angles of a triangle, then $\alpha + \beta + \gamma =$ two right angles.

Proof. Consider the triangle ABC shown in Figure 2.8, with angles α, β, γ at the vertices A, B, C respectively, and draw a parallel to AB through C. Because of the parallel lines, we get alternate angles: another angle α at C

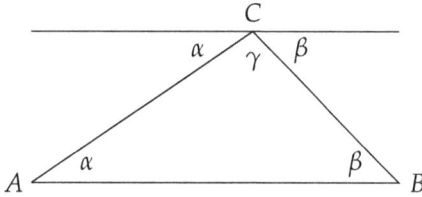

Fig. 2.8   Angles in a triangle

as shown, and another angle β likewise. Together, with the angle γ already at C, these alternate angles give

$$\alpha + \beta + \gamma = \text{ straight angle } = \text{ two right angles,}$$

since a straight angle equals two right angles by definition of the right angle.   Q.E.D.

Now that we know the angle sum of a triangle, we can find the angle sum of any other polygon by dividing it into triangles. For example, a pentagon can be divided into three triangles (Figure 2.9), hence the angle sum of a pentagon equals the angle sum of three triangles; namely, six right angles.

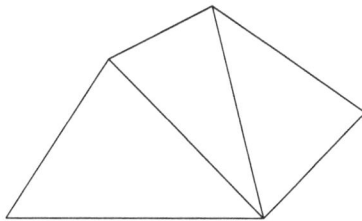

Fig. 2.9   Finding the angle sum of a pentagon

To anticipate something that will come up later, let us call the straight angle π. Yes, this is the same π you already know in connection with the circle, but for now π is only a name for the straight angle. With this naming, the angle sum of any triangle is π and the angle sum of any pentagon is 3π. If we have an equilateral triangle, whose angles are equal by SSS, then each angle equals $\pi/3$. And if we have a **regular pentagon**, whose five angles are equal by definition, then each angle equals $3\pi/5$.

You similarly find (as if you didn't already know) that the angles of a square are $\pi/2$, and the angles of a regular hexagon are $2\pi/3$. These angles make possible something else you may already know: the plane can be **tiled** with copies of the equilateral triangle, or copies of the square, or copies of the regular hexagon. To "tile the plane" with copies of a polygon means to fill the plane with them, without overlapping or leaving gaps. Figure 2.10 shows the tilings by equilateral triangles, squares, and regular hexagons. The tilings by equilateral triangles and hexagons are shown together, in different colours, because of the pretty relationship between them.

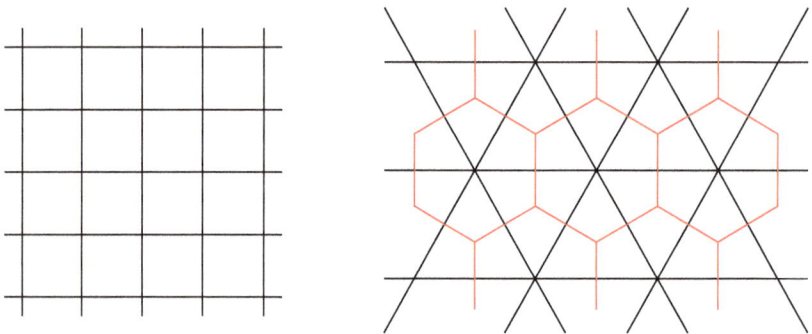

Fig. 2.10   Tiling the plane by regular polygons

Are there any tilings by other regular polygons? Well, no. Pentagons are no good because they leave gaps if you put three of them together at a vertex: the three angles add up to $9\pi/5$, which is less than the 2π needed to fill the space around a point. And if you try to put four pentagons together they will overlap. A regular polygon with more than six sides is also no good. It has angles greater than $2\pi/3$, hence overlapping will occur as soon as three copies of the polygon come together. (We are assuming here that tiles are laid so that edges match, so a vertex of one tile cannot meet another tile in the middle of an edge. But the latter possibility can also be ruled out by considering angles where tiles come together.)

As a remarkable final example of a theorem about angles, let's look at Euclid's Proposition 20 of Book III. Its proof nicely combines Theorem 2 on isosceles triangles with Theorem 3 on the angle sum of a triangle.

Theorem 4. If A and B lie on a circle with centre O, and C is any point on the circle, then the angle ACB is half the angle AOC (and hence is independent of C).

Proof. First look at Figure 2.11, which shows two points A and B on the circle and the angles they create at the centre O and at another point C on the circle. The first step of the proof is to draw the line OC, which divides the angle at C into two parts α and β.

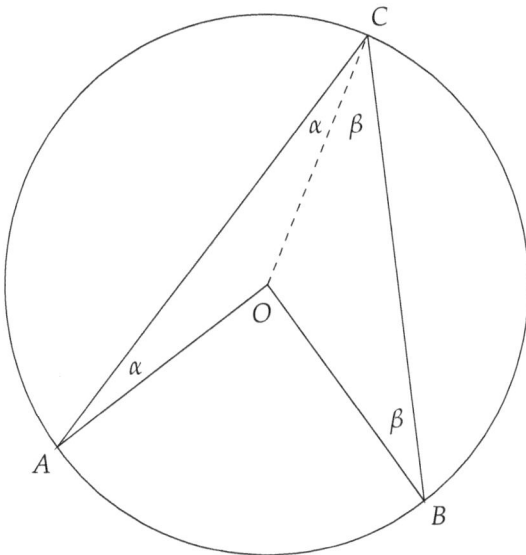

Fig. 2.11   Angles in a circle

Then we notice that the triangles AOC and BOC are both isosceles, because each has two sides that are radii of the circle. Therefore, by Theorem 2, triangle AOC has a second angle α and triangle BOC has a second angle β, as shown.

The second step is to find the third angle in each triangle. Since the angle sum of each triangle is π, the third angle of triangle AOC is $\pi - 2\alpha$ and the third angle of triangle BOC is $\pi - 2\beta$.

Finally, the three angles at O sum to 2π, so the remaining angle at O is

$$2\pi - (\pi - 2\alpha) - (\pi - 2\beta) = 2\alpha + 2\beta = 2(\alpha + \beta),$$

which is twice the angle at C.   Q.E.D.

I should confess that my proof above assumes that C is in a "nice" position, so that the triangles AOC and BOC do not overlap. They would overlap if, for example, the point C was moved clockwise to the 3 o'clock position. Nevertheless, the proof is similar in this case, with just some changes of sign. (Try it!) Another interesting question to think about is: what happens if A and B are at opposite ends of a diameter?

2.5   Some "Unsolved" Greek Problems

Now that you know how to bisect an angle it makes sense to think about cutting an angle into three equal parts using ruler and compass. Then we might try cutting an angle into four equal parts, or more. Four *is* possible, because after bisecting an angle, you can bisect each part. After that you can bisect each of the four equal parts, getting eight, and so on. But this does not help us to get three equal parts of an angle; an arbitrary angle, that is. There is certainly at least *one* angle that can be divided into three equal parts. Can you name one, or more?

Dividing an arbitrary angle into three equal parts, or **trisection of the angle**, was a problem of great interest to the ancient Greeks. They managed to do it with some fancier instruments but it wasn't until the 19th century that mathematicians were able to show that it could not be done by ruler and compass.

A mathematically similar problem was one called **doubling the cube**.

Here the aim was to construct a cube with volume twice the volume of a given cube. More precisely: given the edge length of a cube, construct the edge length of a cube with twice the volume of the given cube. This problem, too, was shown to be impossible in the 19th century. Perhaps surprisingly, these two geometric problems were eventually shown impossible by algebra. At this stage, let me just dangle a hint: trisection and doubling the cube are "cubic" problems, whereas problems solvable by ruler and compass are "quadratic". For more on this, see Section 12.3.

Related to trisection, but more general, was the problem of constructing regular polygons by ruler and compass. As we know, Euclid constructed the regular triangle – the equilateral triangle – in his very first proposition. Later, in Book IV, Proposition 11, he solved the more difficult problem of constructing the regular pentagon. Then he combined the two constructions to construct a regular 15-gon in Book IV, Proposition 16.

You may wonder: what is so special about 15? Probably not much, except that a regular 15-gon is essentially all you can make when you have only the regular 3-gon and 5-gon (the equilateral triangle and regular pentagon). Apart from this, all you can do is double the number of sides, any number of times, by bisecting angles. There were no other options until 1796, when the 19-year-old Carl Friedrich Gauss found a ruler and compass construction of the regular 17-gon. What Gauss actually discovered was that *a regular n-gon is constructible by ruler and compass if n is the product of a power of 2 by distinct primes of the form* $2^{2^k} + 1$.

With this amazing discovery, a Greek geometry problem suddenly became a problem of number theory; namely, which numbers of the form

$$n = 2^{2^k} + 1$$

are prime numbers? If you substitute $k = 0, 1, 2$ in this formula you get the numbers $n = 3, 5, 17$, which are primes. They correspond to the regular n-gons constructed by Euclid and Gauss. It so happens that $k = 3, 4$ give the numbers 257 and 65537, which are also prime. Hence regular polygons with 257 or 65537 sides are also constructible by ruler and compass (but don't try it at home!).

The catch is: we still don't know whether $2^{2^k} + 1$ is a prime for any larger value of k! Thus we don't yet know exactly which regular n-gons are constructible.

The deepest of the ancient Greek problems yet settled was called **squaring the circle**. It required constructing a square with area the same as the area of a given circle. This problem was eventually proved impossible to solve by ruler and compass in 1882, but by highly sophisticated methods beyond the scope of this book. As you may suspect, the problem has to do with the nature of π – viewed now as a *number* and not simply as a name for a straight angle.

However, now is a good time to comment on the term "squaring". As I mentioned in Section 2.2 on ingredients of the *Elements*, Euclid did not use numbers to measure lengths or areas. An area literally *was* an object in the plane, and the standard objects for representing area were squares. Thus, for the Greeks, to "find the area" of a given object in the plane meant finding a square *equal* to it. And how did one decide, without using numbers, when objects were "equal"? I say more about this in the next section, on Pythagoras' theorem.

2.6   Who Discovered Pythagoras' Theorem?

Controversy and Proofs

Pythagoras' Theorem. The square on the hypotenuse of a right-angled triangle equals the sum of the squares on the other two sides (Figure 2.12).

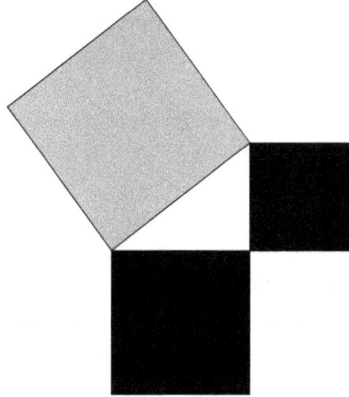

Fig. 2.12   The right-angled triangle

You might think that Pythagoras discovered the theorem named after him, but the story is more complicated than that. The theorem was discovered independently in many places, and some of the discoveries definitely preceded Pythagoras. Also, different *aspects* of the theorem came to light in different places, and at different times. Here are three of them, which turn on different interpretations of the "sum of squares".

(1) Whole numbers a, b, c such that $a^2 + b^2 = c^2$ are called **Pythagorean triples**, but Pythagoras and his followers were not the first to find them. An amazing collection of Pythagorean triples was tabulated in Babylonia somewhere between 1900 BCE and 1600 BCE, more than 1,000 years before Pythagoras. The table has 15 examples, involving triples as large as $(13500, 12709, 18541)$, so the Babylonians must have had a powerful method of finding them. Also, the triples (a, b, c) are listed in decreasing order of b/a, which we would interpret as the slope of the hypotenuse – suggesting that the Babylonians were aware of their interpretation as sides of right-angled triangles.

(2) When "squares" are interpreted literally as geometric squares on the sides of a right-angled triangle, there are simple pictures that explain why the square on the hypotenuse equals the sum of the squares on the other two sides. Chinese and Indian mathematicians discovered some of these, probably before Pythagoras, and we will study one of them below.

(3) When one tries to reconcile the interpretation of a, b, c as whole numbers with their interpretation as line segments, a conflict arises in the case where $a = b$. In this case, as some follower of Pythagoras discovered, there are *no* whole numbers a, b, c such that $a^2 + b^2 = c^2$. In my opinion, this was the most (mathematically) fruitful aspect of Pythagoras' theorem and it was highly influential in the development of Greek mathematics. This was the discovery of **irrationality** mentioned in the outline of the *Elements* at the beginning of this chapter, and explained in the case where $a = b$ at the end of this section.

So now it is time to turn to Euclid.

Euclid has two very different proofs of Pythagoras' theorem, in Books I and VI of the *Elements*. In Book X he also has a formula for Pythagorean triples and an extensive discussion of irrationality, so the *Elements* in their day were a one-stop-shop for all things Pythagorean.

His proof in Book I, Proposition 47, is based on the simplest facts about the area of squares, parallelograms, and triangles, proved by adding and subtracting triangles. Still, the proof is lengthy, demanding that many triangles be proved congruent, and I would like to offer a more memorable alternative. It is based on Figure 2.13, whose origin is unknown, but the idea seems too obvious not to have been thought of a long time ago.

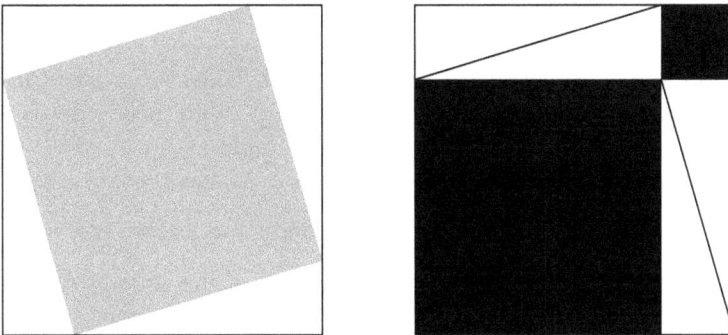

Fig. 2.13   Seeing the Pythagorean theorem

In the square on the left, four copies of the right-angled triangle are arranged so as to leave a grey square in the middle, which is obviously the square on the hypotenuse. In the square on the right the four triangles are arranged so as to leave two black squares, which are obviously the squares on the other two sides. The grey and black regions are *equal* because each equals the big square minus four triangles.

This looks like a proof, but to make this argument as rigorous as Euclid's I should really check that the grey and black regions are squares. This amounts to showing that each has four equal sides and a right angle at each corner. The equal sides follow from equality of the triangles, and the right angles follow from the fact that the angle sum of a triangle is π. For example, each corner of the grey square is part of a straight angle π, the other two parts of which sum to $\pi/2$, because they are the non-right angles of the triangle.

In contrast to the proofs of Pythagoras' theorem, which are numerous and often very clever, the *converse* theorem has essentially only one proof. It was given by Euclid immediately after his first proof of Pythagoras (*Elements*, Book I, Proposition 48). To be concise I will write a^2, b^2, c^2 for the squares on the sides a, b, c of the triangle, and $+$ for the sum of squares, as is usual today.

Converse Pythagoras' Theorem. If a triangle has sides a, b, c such that
$$a^2 + b^2 = c^2,$$
then the sides a and b form a right angle.

Proof. There *is* a triangle with sides a, b, c, where a and b form a right angle. This is because, if we make a triangle with a right angle between sides a and b, its hypotenuse is necessarily c, where $c^2 = a^2 + b^2$, by Pythagoras' theorem.

And there is really *only one* triangle with the sides a, b, c, since all such triangles are congruent, by SSS. So any triangle with these sides is right-angled, with sides a and b forming the right angle.   Q.E.D.

We can write the statements of the theorem and its converse as

(i) if a triangle is right-angled, then the sum of the squares on the two smaller sides is equal to the square on the third side, and

(ii) if the sum of the squares on two sides of a triangle is equal to the square on the third side, then the triangle is a right-angled triangle.

This can be shortened to

> A triangle is right-angled if and only if the sum of the squares on two sides is equal to the square on the third side.

The "if" part of the "if and only if" above is equivalent to statement (ii) and the "only if" is equivalent to statement (i). We can determine which side is the hypotenuse because the hypotenuse is the biggest side of a right-angled triangle.

According to Scott Loomis (1928), there were 370 proofs of Pythagoras' theorem known in the late 1920s. This number has certainly grown and it grew by at least one more in 2023. Its authors were both teenage schoolgirls from New Orleans! It is very unusual for anyone still in school to present a mathematical paper at a professional research conference. But Calcea Johnson and Ne'Kiya Jackson gave their proof at the American Mathematical Society's (AMS) Spring Southeastern Sectional Meeting in Atlanta in March 2023.

You might ask why you would bother to look for a new proof of a more than well-known theorem. Well check out Dawson (2015). He gives eight reasons why you might want to re-prove a theorem.

(1) To remedy perceived gaps or deficiencies in earlier arguments;
(2) To employ reasoning that is simpler, or more perspicuous, than earlier proofs;
(3) To demonstrate the power of different methodologies;
(4) To provide a rational reconstruction (or justification) of historical practices;
(5) To extend a result, or to generalise it to other contexts;
(6) To discover a new route;
(7) Concern for methodological purity;
(8) To provide something analogous to the role of confirmation in the experimental sciences.

The Meaning of Equality

The proof of Pythagoras' theorem based on Figure 2.13 shows that

$$\text{square on the hypotenuse} + \text{four triangles} =$$
$$\text{sum of squares on the other two sides} + \text{four triangles.}$$

Thus, by "subtracting four triangles from each side" we get Pythagoras' theorem:

square on the hypotenuse = sum of squares on the other two sides.

This is a good illustration of the way Euclid establishes "equality of area" using only his **common notions** from the section on Ingredients at the beginning of this chapter. To reiterate, the relevant notions are:

> Things which equal the same thing also equal one another.
> When equals are added to equals, the wholes are equal.
> When equals are subtracted from equals, the remainders are equal.

Implicitly, Euclid is *defining* equality by these notions. It is a complicated way to define equality, but it works, and it avoids the use of numbers to measure area. It was proved much later (in the 19th century) that *any* polygons of equal numerical area can be proved equal by dividing them into a finite number of triangles, then adding or subtracting. But why were the Greeks so scared of numbers in geometry? This goes back to Pythagoras.

Pythagoras and Friends

Pythagoras was probably born on Samos and lived between 570 BCE and 495 BCE, approximately. In about 530 BCE, Pythagoras moved to Croton, southern Italy (part of the bottom of the foot at the southern end of Italy), and set up a school. His students, the Pythagoreans, had a hard monastic type of life. They were sworn to secrecy about everything they did. This presented at least two problems, one of which is how did Pythagoras' theorem get to be known outside his school. Was there a Pythagorean mole?

Incidentally, the school held all things in common and this included their mathematical achievements. So maybe Pythagoras' theorem wasn't discovered by Pythagoras at all, but by one of his students. The same might be said of other things credited to him. For example, the discovery that the morning and the evening star are actually the same (what we call Venus).

A second problem, one that the Pythagoreans would also have wanted to deliberately hide, was their discovery of irrational numbers. Before that, the group had believed that the only numbers were the whole numbers or ratios of whole numbers (= "rational" numbers) and that everything could be explained in terms of them. This belief was supposedly inspired by the role of numbers in musical harmony, a belief then extended to astronomy in the so-called "harmony of the spheres".

Then they proved his theorem and it seemed that if $a = 1 = b$, then $c = \sqrt{2}$ was something they didn't want to know about. $\sqrt{2}$ is **irrational**; that is, not a ratio of whole numbers ("irrational" = not a ratio). In other words, they had discovered numbers that are not **fractions**.

This discovery was very unwelcome to the Pythagoreans, who had hoped to explain the world in terms of whole numbers. There is a legend that one of them leaked the idea of irrational numbers and was drowned in the sea as a punishment.

Here is a proof that $\sqrt{2}$ is irrational, probably similar to the original.

Irrationality of $\sqrt{2}$: $\sqrt{2}$ is not a fraction.

Proof: Suppose that $\sqrt{2}$ is a fraction. Also suppose that a and b are whole numbers with no common divisor and that $\sqrt{2} = a/b$. Then $2b^2 = a^2$. But since the left side is divisible by 2, then a must be divisible by 2. Let $a = 2c$. Then $a^2 = 4c^2$ and hence $b^2 = 2c^2$. This last equation tells us that b is divisible by 2. That means that a and b *do* have a common divisor, contradicting our original assumption. Hence $\sqrt{2}$ is *not* a fraction. Q.E.D.

This kind of proof is also called a **proof by contradiction**. A hypothesis is stated as true. Then the proof produces a contradiction, so the hypothesis has to be wrong. So in fact the opposite of the hypothesis is true. That's how the "$\sqrt{2}$ not a fraction proof" works.

A slight variation of the proof gets the contradiction by "infinite descent". If we don't bother to stipulate that a and b have no common divisor, then of course it will not be a contradiction when we find that a and b have the common divisor 2. However, in that case we get *smaller* whole numbers, $a' = a/2$ and $b' = b/2$ such that $2b'^2 = a'^2$, and we can apply the argument again to show that a' and b' have the common divisor 2. And so on, again and again, giving smaller whole numbers forever. Now *this* is a contradiction, because any descending sequence of whole numbers is finite.

That leads us to look for other irrationals. Are $\sqrt{3}$, $\sqrt{4}$, $\sqrt{5}$, and so on irrational? Are there any more irrational numbers like this? Are any of $\sqrt[3]{3}$, $\sqrt[3]{4}$, $\sqrt[3]{5}$ irrational too? Chapter 3 will reveal a powerful way to answer this question – the **fundamental theorem of arithmetic** – but you might like to try your hand on some particular numbers first.

For more details on Pythagoras and the Pythagoreans you might like to read Wikipedia, Pythagoras.

2.7 Platonic Solids

In this section I want to look at some solids with exceptional symmetry, but before that let's look at some symmetric planar objects: regular polygons. These are polygons that have all sides equal and all interior angles equal. The well-known ones are the equilateral triangle, the square, the regular pentagon and the regular hexagon, already discussed in Section 2.4.

But to the heading of this section, Platonic solids (see Wikipedia, Platonic solid). What are they? Well let's go back to the regular polygons for a start. They are "nice" in some way as they have more symmetry than most polygons, so that is probably why they are of interest to mathematicians (and others). The Platonic solids are in some way the three-dimensional ideal of niceness. (Setting aside the sphere perhaps.) In fact, as you can see below, they borrow some of their niceness from two dimensions.

What is nice about the Platonic solids? Well, it's what you might expect by building on the niceness of the regular polygons. First, every face of each solid is the same regular polygon. Second, the same number of faces meet at a vertex. Plato (with the help of Theaetetus) felt that the solids in Figure 2.14 were the only ones, but Euclid gave the first proof we know of that there are just five of them (in the *Elements*, Book XIII). The solids are the tetrahedron, cube, octahedron, dodecahedron, and icosahedron.[1]

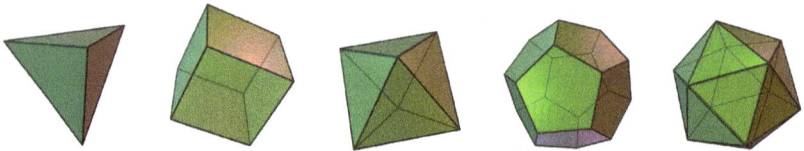

Fig. 2.14   The Platonic solids (see Wikipedia, Platonic solid)

The proof that there are *no other* regular solids is similar to the proof in Section 2.4 that only the equilateral triangle, square, and regular hexagon can tile the plane. You consider the interior angles of each regular polygon, but this time the condition is not that the angle sum at each vertex equals 2π, but that it be less than 2π. However, that's the easy part. Euclid, conscientious as ever, felt compelled to prove that the five Platonic solids

---

[1]As you may guess from their names, these objects are examples of **polyhedra**. The odd one out, the cube, can also be called a hexahedron. However, I will generally use the looser term "solid", and call the regular polyhedra by the traditional name "Platonic solids".

really *exist*. How can you be sure, for example, that 12 pentagons actually fit together exactly to form a dodecahedron? Well, Euclid proved they do.

Plato got his name on the regular solids because he thought that the first four were the shapes of the elements of fire, air, water and earth. What's more, he believed that everything in the universe was made from these elements. The dodecahedron he consigned to the sky where there are 12 constellations. These all appear in Plato's dialogue Timaeus (see Wikipedia, Platonic solid).

We'll return to these solids in later chapters when we discuss the so-called **golden ratio**. Incidentally, the cube is the only one of these solids that will "tile" space.

2.8 Non-Euclidean Geometry

Euclid's *Elements* was the textbook for countless generations of students, including some who became eminent mathematicians. Despite this, no one understood the *Elements* completely – or even very well – until the 19th century. In particular, no one knew whether the parallel axiom was really necessary, nor did anyone notice the flaws such as the faulty "proofs" of SAS and SSS.

A critique of the *Elements* began around 1800, when Gauss and some of his friends began to explore the consequences of supposing that the parallel axiom is false. There are two ways this could happen:

(1) There could be *no* parallels. That means any two lines in the plane would intersect.
(2) There could be *more than one* parallel to a given line through a given point.

At this moment I should remind you that the terms "points", "lines", and "planes" are undefined; they can stand for any objects whatever as long as these objects satisfy Euclid's Postulates. Here is an example.

Let's interpret the "plane" as the surface of a sphere, "points" as points on the sphere, and "lines" as great circles on the sphere. We can also interpret "angles" as angles between great circles. And as before, a "right angle" is half of a straight angle. Then the geometry of the sphere resembles Euclid's geometry quite strongly. Figure 2.15 shows three "lines" on the sphere, which divide the sphere into eight spherical "triangles". These spherical objects satisfy some of Euclid's axioms. For example, Postulate 1 is satisfied, because there is a "line" through any two points. Postulate 3 also holds, because there is a "circle" with any "radius" (meaning any segment of a "line") about any point, and Postulate 4 holds because all "right angles" are equal. Better still, the axioms that Euclid failed to include, SAS and SSS, are also satisfied.

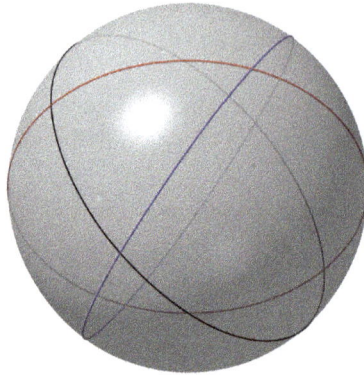

Fig. 2.15   Some "lines" and "triangles" on the sphere

However, the parallel axiom is obviously *not* satisfied on the sphere, because any two "lines" meet. Postulate 2, saying that any line segment can be extended, also fails because great circles are finite. It turns out that this is a feature of all interpretations in which there are no parallels. So if we hope to satisfy all of Euclid's axioms *except* the parallel axiom, we need to look for interpretations with more than one parallel to a given "line" through a given "point".

Such interpretations, or **non-Euclidean planes**, are hard to find, though the sphere is kind of a clue. The "lines" on the sphere are "shortest possible" in the sense that the shortest path between any two points is a segment of a great circle. This suggests studying "shortest possible" paths, called **geodesics**, on other surfaces and taking them to be the "lines" in a new geometry. And in fact there *are* surfaces whose geodesics satisfy all of Euclid's axioms except the parallel axiom. They were found by Eugenio Beltrami in 1868. Unfortunately, they do not exist in ordinary three-dimensional space, only in a more abstract realm.

The good news is that these abstract planes can be mapped into the ordinary plane, though the image is necessarily somewhat distorted. This, too, is something we are used to with the sphere – we know that maps in an atlas give a somewhat distorted view of the Earth, but we still can read them. One image of a non-Euclidean plane is shown in Figure 2.16.

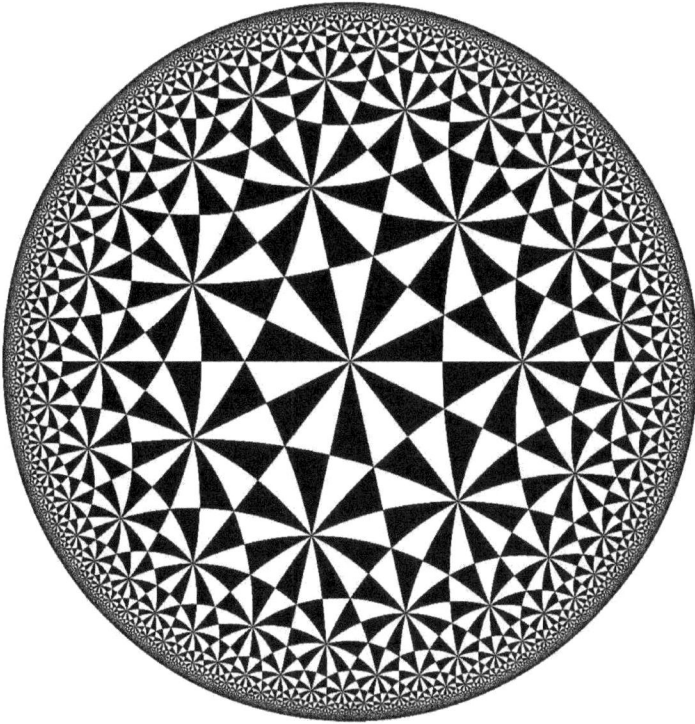

Fig. 2.16   A tiling of the non-Euclidean plane (Wikipedia, (2,3,7) triangle group)

The "points" of this non-Euclidean plane are points inside the boundary circle, and its "lines" are circular arcs perpendicular to the boundary circle. Obviously, straightness and length are distorted in this picture. But, happily, angles are not. The "angles" in each triangle are the actual angles at which the circular arcs meet. Each triangle has angles $\pi/2$, $\pi/3$, and $\pi/7$, which you can check by counting the numbers of triangles that meet at each vertex (either four, or six, or fourteen). In non-Euclidean geometry, triangles with the same angles are *congruent* so we are looking at a *tiling of the non-Euclidean plane* by congruent triangles.

Among other things this shows that "lines" in non-Euclidean geometry have infinite length, because each "line" passes through infinitely many triangles on its way to the boundary. So Euclid's Postulate 2 is satisfied.

But the parallel axiom fails (hooray!) because you can see many "lines" that do not meet the "line" down the centre of the disc, even though they pass through the same "point" outside it. (Also, the angle sum of each triangle is less than π.)

Chapter 3

Number Theory

3.1 Number

In this chapter I begin with some background information on concepts of number theory. "Numbers" for Euclid meant the positive whole numbers 1, 2, 3, 4, 5, …, but to explain some of his results it will help to bring in zero and the negative integers as well.

Before I get started it is worth noting that Euclid does all of his theory of numbers representing numbers as lengths of line segments, and using geometric terms such as "measures" where we would say "divides". This makes his number theory look superficially like geometry, but it really is about arithmetic, and it will be easier to understand if we use the usual language of arithmetic.

Primes

A whole number, other than 1, is called a **prime number**, if it is divisible only by 1 and itself. Now you would think that 1 ought to be a prime number, but it isn't. We'll say more about this soon (in Section 3.4 on the Fundamental Theorem of Arithmetic).

Prime numbers interest people for at least two reasons. First, every number can be written as a product of prime numbers. What's more this prime decomposition of a number only ever uses the same prime numbers. To see what we mean, look at 1200. This can be written as $2^4 \times 3 \times 5^2$. But it can't be written using any other primes, and the numbers of twos, threes and fives are always the same: four twos have to be included, one three, and two fives have to be there too. This is an instance of a theorem saying that every number can be written in only one way as a product of primes (see Section 3.4).

The second reason for the interest in prime numbers is that they are easy to find, hard to get, but very valuable. Eratosthenes showed that they are easy to get. You just use his sieve and they sort themselves out. At least it sounds easy until you get up to the multiples of 863, say. It's easy moving on 2 at a time and taking out all of the multiples of 2. And it's not too hard to move on by 11 to remove the multiples of 11. But counting on from 863 isn't easy unless you have a computer handy. Once you get to numbers with hundreds of digits, even computers get a little exhausted. Yet large primes are valuable because of the RSA cryptosystem, used extensively in business and on the internet (see Section 13.4).

People have tried to find prime number formulas. I'm thinking of things like the Mersenne primes that satisfy the formula $2^n - 1$ (Wikipedia, Mersenne prime). You don't get a prime for every n, for instance try $n = 4$, but every now and again you hit on one. At the time of writing, $n = 82,589,933$ will give you the largest known Mersenne prime. There are almost certainly more of these things to be found, but for numbers so large it takes a while to check their primeness.

The difficulty with primes is that we don't have a formula that will hit a prime every time. The formula $x^2 + x + 41$ is impressive because it gives a prime for every x less than 40, but it obviously fails for $x = 41$, and it's hit or miss after that. Also, the primes that a quadratic gives are so small that they are not worth too much.

Infinite Descent

Euclid didn't use the term infinite descent, but he knew how to use it. We'll hear more about the idea in Fermat (Part 4) when we talk about Fermat's escapades and Pascal's use of an equivalent principle called **induction**. So what is infinite descent?

Infinite descent is a method of proof that works by first assuming that something holds and then showing that, whenever it holds, it also holds for a *smaller* (positive) whole number. Since positive whole numbers cannot decrease forever, this is a contradiction (Euclid said "which is impossible in numbers"). Hence our assumption is false. We used this method once before, in Section 2.6, to prove irrationality of $\sqrt{2}$. The *Elements*, Book VII, Proposition 31, is another example showing that Euclid certainly knew the idea.

Book VII, Proposition 31: Any number has a prime divisor.

Proof: Let a be any number. If a is not prime there is a $b < a$ that divides a. If b is prime then the proposition is proved. If b is not prime, it has a smaller divisor, c, say. And so on. Proceeding in this way, we get a descending sequence $a > b > c > \cdots$, which cannot continue indefinitely. Therefore, at some stage we meet a divisor that is prime. Q.E.D.

This proof can be viewed as a descent, from a hypothetical number with *no* prime factors, to ever-smaller numbers with the same property, which is impossible.

In the next section we will look at a famous algorithm due to Euclid, which relies on infinite descent in order to work. The algorithm produces smaller numbers at each step until they just have to stop. And when the algorithm halts, we simply read off the answer.

3.2 The Euclidean Algorithm

The sieve of Eratosthenes shows that we can produce very many prime numbers, as long as we don't ask for primes with too many digits. In fact, it can be hard to find even one prime divisor of a number with hundreds of digits. Yet when it comes to finding a *common* divisor of two large numbers, there is a fast method. This method is due to Euclid and it is called the **Euclidean algorithm**. He describes it very concisely in Proposition 1 of Book VII of the *Elements*: start with any two numbers and continually subtract the lesser from the greater.

Well, maybe that's a little *too* concise, so let me illustrate the algorithm with an example. Take the numbers 15 and 6. Subtracting the lesser, 6, from the greater, 15, we get

$$15 - 6 = 9.$$

We now have a new pair of numbers, 6 and 9 (which is what remains of 15 after 6 is subtracted). 6 is still the lesser, so we subtract it again, from 9:

$$9 - 6 = 3,$$

getting the new pair of numbers 6 and 3. The lesser is 3, so we subtract it from 6:

$$6 - 3 = 3,$$

getting the new pair of numbers 3 and 3. Neither is the lesser, so the algorithm halts with the answer 3, which is indeed the greatest common divisor of 15 and 6.

We notice that 3 is actually the greatest common divisor of *each* pair produced. This, in fact, is why the algorithm works. If d is a common divisor of two numbers $a = da'$ and $b = db'$, then d also divides their difference $a - b = d(a' - b')$. So, by continually subtracting, we produce smaller and smaller numbers with the *same* common divisors – until we finally have a pair of equal numbers, each of which is necessarily the *greatest* common divisor, since it contains all the common divisors.

Euclid was content to use repeated subtraction in his algorithm, and it is easier to understand, but it is a slow method when one number is much larger than the other. For example, if we start with the numbers 51 and 6 we will have to subtract 6 eight times from 51 before we get a remainder (namely, 3) that is smaller than 6. Under these circumstances, a better idea is to use **division with remainder.**

That is, given that the lesser number is 6, we divide 6 into 51 and find that it goes eight times, with remainder 3:

$$51 = 8 \times 6 + 3,$$

so we can jump straight to the new pair of numbers, 6 (the "divisor") and 3 (the "remainder"), after which the algorithm does another division with remainder. This form of the Euclidean algorithm produces exactly the same result as the subtractive form because division with remainder produces the same result as repeated subtraction. But it is more efficient because division is faster than repeated subtraction. In fact, with the help of a computer, the Euclidean algorithm can easily find the greatest common divisor of numbers with hundreds of digits, because the number of divisions is roughly proportional to the number of digits.

For proving facts *about* the Euclidean algorithm, however, it is easier to work with repeated subtraction, since subtraction is simpler than division with remainder. One such fact is known as **Bézout's identity.**

Bézout's identity. When something is named after Bézout (18th-century French mathematician) that's a pretty big clue that it wasn't discovered by Euclid. Actually, it wasn't discovered by Bézout either, since it was known to Bachet (17th-century French mathematician) more than a century earlier. Nevertheless, Bézout's identity gives a simple way to explain something that *was* known to Euclid, so let's get into it.

From now on, let's abbreviate the greatest common divisor of numbers a and b by $\gcd(a, b)$. For example, $\gcd(15, 6) = 3$. A surprising fact about $\gcd(a, b)$ is that it can always be written as a combination $ma + nb$ of the numbers a and b for integers m and n. This fact is the so-called "Bézout's identity". An example is

$$\gcd(15,6) = 3 = 1 \times 15 - 2 \times 6.$$

Well, we already knew that. Let's try a harder example: $\gcd(13,8) = 1$. Can you find a combination of 13 and 8 that equals 1? It will take you a moment, because the simplest solution of

$$1 = 13m + 8n$$

is $m = -3$ and $n = 5$. If that's too easy, try finding m and n so that

$$\gcd(34,21) = 1 = 34m + 21n.$$

It's not obvious is it? Nevertheless, it is easy to see why there *exist* integers m, n with $\gcd(a,b) = ma + nb$ if we look again at the Euclidean algorithm. Remember, the Euclidean algorithm starts with the numbers a and b and does nothing except subtracting one number from another. It follows that *every number produced by the Euclidean algorithm is of the form $ma + nb$ for some integers m and n.*

Why? Well, this is certainly true at the beginning, because

$$a = 1 \times a + 0 \times b \quad \text{and} \quad b = 0 \times a + 1 \times b.$$

And if the numbers at any stage of the algorithm are both of the form $ma + nb$, then so are the numbers at the next stage, since the difference of two numbers of this form is again of the form $ma + nb$. In particular, this is true at the *last* stage of the algorithm, when each number equals $\gcd(a,b)$.

At this point, the sort of question that a mathematician might ask is "are m and n unique?" Let's look for more solutions of $13m + 8n = 1$, beyond the solution $m = -3$, $n = 5$ already discovered. One way to get another solution would be to add a solution of $13s + 8t = 0$, getting $m = -3 + s$ and $n = 5 + t$, because in that case

$$13(-3 + s) + 8(5 + t) = (-13 \times 3 + 8 \times 5) + (13s + 8t) = 1 + 0 = 1.$$

And $13s + 8t = 0$ is easy to solve! For example let $s = -8$ and $t = 13$. More generally, we can take $s = -8u$ and $t = 13u$ for any number u, which gives *infinitely many* solutions of $13s + 8t = 0$. These solutions in turn give infinitely many solutions $m = -3 - 8u$ and $n = 5 + 13u$ of the original equation $13m + 8n = 1$.

In the same way, we can find infinitely many ways to express $\gcd(a,b)$ in the form $ma + nb$, for any numbers a and b.

A Stamp Problem

In the olden days when stamps were commonly licked and not just printed on your envelope, a post office somewhere suddenly discovered it had only 3c and 7c stamps. With only those amounts available, the postmistress wondered what amounts of postage she could give her clients. So she sat down and worked it out using Table 1, where Y means that it is possible and N that it's not.

Table 1: Experimenting with 3c and 7c stamps.

1	2	3	4	5	6	7	8	9	10	11	12	13	14	15	16
N	N	Y	N	N	Y	Y	N	Y	Y	N	Y	Y	Y	Y	Y

It looks as if it might be possible to get all the amounts from 12c up, but there were a few denominations under 12c that won't work. Is it possible that from 12c onward the sums can all be made up?

The easy solution to this is to note that she could certainly get 12c, 13c, and 14c. But then she could add a 3c stamp to each of those values to get 15c, 16c, and 17c. Of course, from here she could get anything she wanted by adding on more 3c stamps.

At first sight this problem seems a little like the problem of expressing $\gcd(3,7)$, which is 1, in the form $3m + 7n$. After all, if $1 = 3m + 7n$, then $k = 3mk + 7nk$, so we could express any number k as a combination of 3c and 7c stamps. The problem is, we may need a *negative* number of one type of stamp – and post offices cannot supply a negative number of stamps! We see here one of the reasons for inventing negative numbers: they can make problems easier to solve (even problems about positive numbers, as we will see in Section 3.4).

On the other hand, you might say that sticking to positive numbers makes for more interesting problems, and the stamp problem is one of them.

3.3 Infinitely Many Primes

Euclid often gave proofs for an example or two and then assumed, or left the reader to see, why the examples gave the method of proof generally. For example, let's prove that there are an infinite number of primes. I'm first going to believe that you think there are only a few primes, say 2, 3, and 5. (In the next section I'll explain why 1 is excluded from the exclusive club of primes.) My reply is to write down $A = 2 \times 3 \times 5 + 1$. Now the first thing to notice is that none of 2, 3 and 5 divides A, because they each leave remainder 1. So, A is either a prime, or it has a divisor bigger than 5. Ah, $A = 31$, which certainly is a prime.

"Oh," you say, "then the only primes are 2, 3, 5, 31!"

To which I reply that you should look at $B = 2 \times 3 \times 5 \times 31 + 1$. Again 2, 3, 5, 31 do not divide B. So, either B is a prime or it has a prime divisor that is not among 2, 3, 5, 31. Um, $B = 931 = 7^2 \times 19$, so we've collected two more primes here, 7 and 19.

At this point I expect that you will give up and say that there is an infinite number of primes. Of course, you are right. You can see how the argument goes: either the capital letter is prime or it has a prime divisor. Because, you know (from the discussion of infinite descent in Section 3.1) that Euclid had already proved every number greater than 1 has a prime divisor. End of story.

A modern version of Euclid's proof, with the help of subscripts, would go like this. Suppose we have some prime numbers $p_1, p_2, p_3, \ldots, p_n$. Let
$$C = p_1 p_2 p_3 \cdots p_n + 1.$$
Then by the previous argument, either C is a prime or C is divisible by a prime not among $p_1, p_2, p_3, \ldots, p_n$. So, given any collection of primes, we can always find one more; hence there are an infinite number of them.

It is worth dwelling on this argument a little longer. Notice that Euclid has managed to get by without knowing anything about the way primes are distributed among the other numbers. He needs to know only that each number has a prime divisor. His proof is not only the first and simplest proof that there are infinitely many primes, it was the *only* known proof until the 18th century. (Just recently, ChatGPT was asked to produce a proof that there are infinitely many prime numbers, and yep, it actually referred to Euclid's proof.)

Actually, there are many proofs in mathematics where the general case follows rather easily on the model of specific cases, as in the proof above. But whatever, mathematicians need to prove the general case.

3.4 Fundamental Theorem of Arithmetic

Unique Prime Factorisation

The point of this section is to show that any number is the product of primes in a unique way. Remember I talked about this near the start of

Section 3.1. Take 54 for example. Now $54 = 2 \times 27 = 2 \times 3^3$. Of course, you can write it as $3 \times 2 \times 3 \times 3$, or in many other ways, but 54 can only be made up by multiplying together one 2 and three 3s. No 5 or 7 or any bigger prime would even think about appearing in the factorisation of 54.

Now here is the beginning of the demise of the number 1 as a paid up member of the prime family. It was very sad to be booted out, especially as it had wanted to be used endlessly. On the first hand, 1 had sued the Mathematicians Universe Club (Keos) (MUC(K)) for damages but they have ruled it out on a technicality. The technicality: fecundity!!!

The argument for the prosecution went like this. "Yes we understand the defence. A prime is a number divisible only by 1 and itself. The defendant clearly satisfies that criterion. If we allow 1 to be a prime though, then 54 could be written as $2 \times 3^3 \times 1$ or $2 \times 1^5 \times 3^3$ or $1 \times 2 \times 3^3 \times 1^{78}$ or even 2×3^3. These are all different! M'lord you can't have multiple ways to write a number if you want it to be written uniquely!! I therefore ask for 1 to be no longer prime. Then we would avoid this mess of 1 occurring any time it liked and as often as it liked."

Now the jury was stacked with mathematicians and they just decided that was that for 1 as a prime. Consequently, a prime is now defined as a number, *other than* 1, that is only divisible by itself and 1. Poor 1 has never recovered.

Note that Pluto got kicked out of the Planet club because he wasn't quite big enough. Pluto is now called a dwarf planet. It's just not fair. Pluto's problem was that it was just a big rock and out where Pluto lives there are lots more big rocks about his size. The astronomers didn't like to have a fecundity mess either (though perhaps they are more lenient than mathematicians and would allow 1 to be called a "dwarf prime").

Fine, now that I've got that sad story off my chest I have to prove that there is indeed a unique factorisation theorem for primes and that it was essentially proved in the *Elements*. The latter is easy. You only have to look at Book VII Propositions 30, 31 and 32 along with Book IX, Proposition 14. But that makes for hard reading because these things are written in an old-fashioned way. So I'll use another approach that is founded on Euclid's work and Bézout's identity.

The proof of the Fundamental Theorem of Arithmetic involves a type of proposition I should mention, a **lemma**. Usually a "small" theorem that is used to prove a big one, a lemma typically has more general value. Here the lemma is small but it does all the hard work for the theorem.

Euclid's Lemma: Let p be a prime that divides $n = ab$, where a and b are some whole numbers. Then p divides at least one of a or b.

Proof: If p divides a, then we are finished. So we assume that p does not divide a. In that case only 1 divides them both and hence $\gcd(p, a) = 1$.

By Bézout's identity, there exist r and s such that $1 = rp + sa$ hence $b = rpb + sab$. Now p divides both p and ab on the right side of the equation. Hence p must divide b on the left side.   Q.E.D.

Fundamental Theorem of Arithmetic: Every positive number n greater than 1 can be written as a product of primes in only one way.

First let me say that obviously you can write 6 as 2×3 or 3×2, which appear different to most of us. However, we ignore the order of factors. When we say "in only one way", we mean that any prime factorisation contains the same primes, and each prime occurs the same number of times. Mathematicians often say prime factorisation is unique "up to order".

Proof: Let n be the smallest number with two prime factorisations. So

$$n = p_1 p_2 p_3 \cdots p_i = q_1 q_2 q_3 \cdots q_j.$$

Where $p_1, p_2, \ldots, p_i, q_1, q_2, \ldots, q_j$ are prime numbers. Since p_1 is a prime that divides n it also divides $q_1 q_2 \cdots q_j$. By Euclid's Lemma, p_1 divides one of the primes $q_1, q_2, \ldots, q_j$. Without loss of generality we can assume that p_1 divides q_1. Since q_1 is prime, $p_1 = q_1$. Then by cancelling we can see that $m = p_2 p_3 \cdots p_i = q_2 q_3 \cdots q_j$, which again are different factorisations, since we have merely cancelled identical factors from the different factorisations of n. But m is smaller than n and n was supposed to be the smallest number with more than one prime factorisation. This contradiction shows that all whole numbers have a unique prime factorisation.
                                                                    Q.E.D.

This factorisation in general is written as $n = p_1^{a_1} p_2^{a_2} \cdots p_n^{a_n}$, where the primes are usually written in increasing order. For example, $2^3 3^4 5^2$ is the unique prime factorisation of 16200.

It's worth noting that other areas of mathematics have their own Fundamental Theorems. I've given two below.

Fundamental Theorem of Algebra. Every polynomial equation of degree n and with complex coefficients has n complex number solutions.

Fundamental Theorem of Calculus. If $F(x) = \int_a^x f(t)dt$ then $F'(x) = f(x)$.

This last theorem needs some condition on the function involved that I'd rather not include here because of its complexity.

3.5   Irrational Numbers

In Section 2.6 I promised that the fundamental theorem of arithmetic would provide a powerful new method of proving irrationality – for numbers such as $\sqrt{3}$, $\sqrt{5}$, $\sqrt[3]{3}$, $\sqrt[3]{4}$, $\sqrt[3]{5}$ – so now it is time to deliver on that promise. But first, in case you are wondering whether irrational numbers have any business being in a chapter about whole numbers, let me assure you that these ones do. The reason is that asking whether $\sqrt{3}$ is rational or not is the same as asking whether or not the equation $3b^2 = a^2$ has a solution with *whole numbers* a and b, and there are similar equations for the other numbers above. For example $\sqrt{5} = a/b$ if and only if $5b^2 = a^2$.

To show that $3b^2 = a^2$ does not have a whole number solution, suppose (hypothetically) that it does, then consider the prime factorisations of a and b.

Now, whatever the prime factorisation of a (or b) may be, each prime appears *twice as often* in the factorisation of a^2 (or b^2). So, if 3 appears in the prime factorisation of b, it appears an even number of times in the factorisation of b^2, and hence an *odd* number of times in the prime factorisation of $3b^2$ – since the prime factorisation is unique.

On the other hand, 3 appears an even number of times in the prime factorisation of a^2, so the hypothetical equation $3b^2 = a^2$ is impossible, by unique prime factorisation again. In other words, $\sqrt{3}$ is irrational.

There is a similar proof that $\sqrt{5}$ is irrational, by considering how often the prime 5 appears on each side of the hypothetical equation $5b^2 = a^2$. And a slight elaboration proves $\sqrt{n}$ is irrational for any whole number n that is not a perfect square. Try your luck with $n = 6 = 2 \times 3$.

But what about cube roots, such as $\sqrt[3]{2}$? Here we are looking at the hypothetical equation $2b^3 = a^3$, and we would like to show that this equation contradicts unique prime factorisation because the prime factor 2 occurs a different number of times on each side of the equation. Well, does it?

No matter how many times 2 appears in the prime factorisation of b, it appears *three times as often* in the prime factorisation of b^3, so the number of its occurrences in $2b^3$ is a *multiple of 3, plus 1*. On the other hand, the number of times 2 appears as a prime factor of a^3 is simply a multiple of 3, so $3b^3 \neq a^3$, by unique prime factorisation. Therefore, $\sqrt[3]{2}$ is irrational.

You should now be able to use unique prime factorisation to prove the irrationality of virtually any root of a whole number, as long as the root is not "obviously" rational. A similar method also works for certain logarithms; for example, $\log_{10} 2$ is irrational. Here is a hint.

$$\log_{10} 2 = \frac{a}{b} \quad \text{if and only if} \quad 10^{a/b} = 2 \quad \text{if and only if} \quad 10^a = 2^b.$$

Approximating $\sqrt{2}$

After being shocked to discover that $\sqrt{2}$ is not a ratio of whole numbers, the Pythagoreans did not give up trying to understand it. Instead, they looked for fractions b/a that "approach" $\sqrt{2}$ in a natural way, and they came up with pairs (b, a) they called **side and diagonal** numbers. The reason for the name is that a rectangle with one side a and diagonal b "approaches" the shape of a square because b/a "approaches" $\sqrt{2}$.

The Pythagoreans found an interesting way to generate side and diagonal numbers, which is an example (possibly the first) of a **recurrence relation** – a process in which each new pair of numbers comes from the previous pair by a certain rule.

We start with the pair $(b_1, a_1) = (3, 2)$ and produce each new pair (b_{n+1}, a_{n+1}) from the previous pair (b_n, a_n) by the rules

$$b_{n+1} = b_n + 2a_n$$
$$a_{n+1} = b_n + a_n. \tag{*}$$

Thus the pair $(3, 2)$ produces the pair $(3 + 2 \times 2, 3 + 2) = (7, 5)$, which in turn produces the pair $(7 + 2 \times 5, 7 + 5) = (17, 12)$, and so on. Thus we get the sequence of pairs

$$(3, 2), \ (7, 5), \ (17, 12), \ (41, 29), \ (99, 70), \ (239, 169), \ (577, 408), \ \dots$$

The fractions corresponding to these pairs approach $\sqrt{2}$ at an impressively rapid rate, as the following decimals show.

$$3/2 = 1.5$$
$$7/5 = 1.4$$
$$17/12 = 1.416\cdots$$
$$41/29 = 1.4137\cdots$$
$$99/70 = 1.41428\cdots$$
$$239/169 = 1.414201\cdots$$
$$577/408 = 1.4142156\cdots$$

$$\vdots$$

$$\sqrt{2} = 1.41421356237\cdots$$

Apparently, each new fraction in the sequence gives about one more decimal place towards the exact value of $\sqrt{2}$. Why is this happening?

Certainly, the numbers b_n and a_n increase rapidly, because the recurrence relations (*) force them to more than double at each step. That's part of the explanation. The other part is that, for each n, it can be shown that

$$b_n^2 - 2a_n^2 = \pm 1.$$

This equation means that the difference between b_n/a_n and $\sqrt{2}$ decreases rapidly. We can see by dividing through by a_n^2 that

$$\frac{b_n^2}{a_n^2} - 2 = \pm\frac{1}{a_n^2},$$

which says that the difference between 2 and b_n^2/a_n^2 decreases *very* rapidly. I hope this convinces you that the difference between $\sqrt{2}$ and b_n/a_n also decreases rapidly. In fact, a little more algebra will show that this difference is even smaller than $1/a_n^2$.

In the next part of the book I will show you some more general ways of calculating irrational square roots.

The Pell Equation

I wrote ± 1 on the right-hand side of $b_n^2 - 2a_n^2 = \pm 1$ because in fact we get $+1$ when n is odd and -1 when n is even. For example, $b_1 = 3$ and $a_1 = 2$ give

$$3^2 - 2 \times 2^2 = 1.$$

Thus the odd-numbered pairs (b_n, a_n) are solutions of the equation

$$x^2 - 2y^2 = 1.$$

This equation is a special case of the so-called **Pell equation** $x^2 - Ny^2 = 1$, where N is a nonsquare whole number.

Pell equations have arisen many times, and in many places, in the history of mathematics, as we will see later in this book. From the case of $\sqrt{2}$ you might expect the general Pell equation $x^2 - Ny^2 = 1$ to give us fractions that rapidly approximate $\sqrt{N}$. But before that, we need to know whether the equation has any solution at all. This turns out to be a very interesting problem – one that was not understood in Euclid's time. To see what happened to the Pell equation next, look at Section 8.5.

3.6 Perfect Numbers

Now I want to look at a mathematician's plaything. As far as I can tell there is no application of perfect numbers, but they are interesting because we don't yet know all there is to know about the topic. And that sort of thing urges pure mathematicians on. Anyway, I'll tell you what perfect numbers are, what Euclid did with them, and what we know about them. As you will see, this topic touches on some of Euclid's finest work, the summation of **geometric series**, so first I would like to give some background about geometric series and their history.

Going to Infinity

Zeno of Elea was a Greek philosopher who lived about a century before Euclid. We know very little about him, but his name is famous in connection with some so-called *paradoxes of the infinite* (see Wikipedia, Zeno's paradoxes). In fact we know of his paradoxes only through Aristotle, who discusses them in his *Physics* only to debunk them, so Zeno's own take on the paradoxes is not known. But no matter, they are a great springboard into the mathematical study of infinity.

Zeno's first paradox is known as the **paradox of the dichotomy**, and it claims that motion is impossible. Because in order to get anywhere one must first get half way there, and before that one quarter of the way there, and before that one eighth of the way there, and so on – so infinitely many events have to take place in a finite time, which is impossible. Or is it? Aristotle says, and we agree with him, that it obviously *is* possible, and indeed the infinitely many places one must visit simply correspond to the infinitely many times at which these places are visited.

More interesting, from the mathematical point of view, are the infinitely many distances that must be covered in order to complete the whole journey. If we take the whole journey to be of length 1, we have broken it into parts of length $1/2, 1/4, 1/8$, and so on, which leads to the equation

$$1 = \frac{1}{2} + \frac{1}{4} + \frac{1}{8} + \cdots,$$

so we have summed the infinite series of terms on the right. This series, each term of which is half the one before, is called an **infinite geometric series**. Thus Zeno's first "paradox" can be viewed as not a paradox at all, but simply the summation of a geometric series.

More difficult geometric series occur in Euclid and also in the work of Archimedes. For example, Euclid finds the volume of a tetrahedron by dividing it into infinitely many pieces, each one a quarter of the one before, so that the total volume is

$$\frac{1}{4} + \frac{1}{4^2} + \frac{1}{4^3} + \cdots.$$

The sum of this series happens to be $1/3$, which leads to the result that the volume of a tetrahedron equals $1/3$ of its base area times its height. I mention this result in case you are wondering what is "geometric" about geometric series. Not a lot, really, but situations where there is a constant proportion between one thing and the next seem to arise more commonly in geometry.

Coming Back from Infinity

Some people think that Zeno's paradoxes were intended to warn mathematicians about the dangers of infinity. After Zeno's time we see Greek mathematicians avoiding infinity where they could, and treating it very carefully when it was unavoidable. Their methods of dealing with infinity became useful after calculus was discovered in the 17th century, and mathematicians began to boldly go to infinity in all directions. For example, to deal with the infinite geometric series

$$a + a^2 + a^3 + \cdots$$

they would replace it by the **finite geometric series**

$$a + a^2 + a^3 + \cdots + a^n,$$

and observe what happens as n increases. Both Euclid and Archimedes avoided infinity in this way.

However, if you are not scared of infinity, the infinite geometric series is actually *nicer* than the finite one; things are simpler when there is "no end". Let me explain the difference in the case where $a = 1/2$, so each term in the series is one half of the one before.

If we are talking, like Zeno, about repeatedly halving some continuous quantity, such as a line segment or a time interval, then we can keep halving indefinitely, as Figure 3.1 suggests. We first cut off the left half, then the next quarter, then the next eighth, and so on. The result is that any point inside the interval eventually falls to the left of a cut, from which we conclude that

$$1 = \frac{1}{2} + \frac{1}{4} + \frac{1}{8} + \frac{1}{16} + \cdots,$$

if we let 1 be the length of the interval.

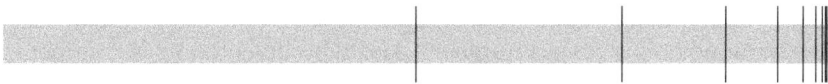

Fig. 3.1   Repeatedly halving a continuous quantity

But if we are halving a *discrete* quantity, such as a whole number N and we insist that the halves also be whole numbers, then the process must end in a finite number of steps. The best-case scenario is when N is a power of 2, in which case we finally halve 2 into $1 + 1$ before we have to stop. Figure 3.2 shows what happens when $N = 2^7 = 128$, in which case we can halve seven times before reaching two ones.

Fig. 3.2   Repeatedly halving a discrete quantity

A thing very like this happens in a tennis tournament such as the Australian Open (men's or women's). Initially there are $128 = 2^7$ players; exactly half the players in each round lose; eventually only two players remain, and the winner of their match is the winner of the tournament. Thus half the players lose in the first round; another quarter lose in the second round; another eighth lose in the third round; until all but one player has lost. Thus the number of players who lose is

$$2^6 + 2^5 + 2^4 + 2^3 + 2^2 + 2^1 + 1,$$

which must equal $2^7 - 1$, the total number of players minus the winner (who is shown as the black dot in Figure 3.2).

I hope you can see what happens when we start with any power of 2. In that case $N = 2^n$ for some n and we get

$$2^{n-1} + 2^{n-2} + \cdots + 2^2 + 2^1 + 1 = 2^n - 1,$$

or, writing the geometric series on the left in increasing order:

$$1 + 2 + 2^2 + \cdots + 2^{n-2} + 2^{n-1} = 2^n - 1.$$

Euclid actually proves a harder result, finding the sum $1 + a + a^2 + \cdots + a^n$ for any numbers a and n, but he needs only the case $a = 2$ to find perfect numbers.

What Does This Have to Do with Perfect Numbers?

Fine, now what is a perfect number? And who invented it? A number is perfect if it is the sum of all of its divisors except the number itself (*proper divisors*). Sometimes these divisors are called *parts*, or *aliquot parts*, so a number is "perfect" if it is the "sum of its parts". For example, 6 and 28 are perfect because

$$6 = 1 + 2 + 3 \text{ and } 1, 2, \text{ and } 3, \text{ are the proper divisors of } 6$$
$$28 = 1 + 2 + 4 + 7 + 14 \text{ and } 1, 2, 4, 7, 14 \text{ are the proper divisors of } 28.$$

You'll be surprised to know that, as far as we know, Euclid was the first person to play with these. With the perfect numbers he knew, he might have noticed that

$$6 \text{ has divisors } 1, 2 \text{ and } 3 \text{ and their sum is}$$
$$(1+2) + 3 = (2^2 - 1) + 2^2 - 1 = 2(2^2 - 1), \text{ while}$$

$$28 \text{ has divisors } 1, 2, 4, 7, 14 \text{ and their sum is}$$
$$(1+2+4) + 7(1+2) = (2^3 - 1) + 3(2^3 - 1) = 2^2(2^3 - 1).$$

The next two perfect numbers are 496 and 8128 and they involve powers of 2 in the same way.

This may have been the trigger to Euclid guessing/conjecturing, that if $1 + 2 + 2^2 + \cdots + 2^{n-1} = 2^n - 1$ is a prime, then

$$2^{n-1}(2^n - 1) \text{ is perfect.}$$

Euclid actually said, but in Greek, Book IX Proposition 36.

> If as many numbers as we please beginning from a unit be set out continuously in double proportion, until the sum of all becomes a prime, and if the sum multiplied into the last make some number, the product will be perfect.

This is known as Euclid's Formula and here is a proof based on the way I played with 6 and 28 above.

Euclid's Theorem: If $2^n - 1$ is a prime then $E = 2^{n-1}(2^n - 1)$ is a perfect number.

Proof: E is the product $2^{n-1}p$ of a power of 2 and the prime $p = 2^n - 1$, so the proper divisors of E are (thanks to unique prime factorisation)

$$1, 2, 2^2, \ldots, 2^{n-1}; p, 2p, 2^2 p, \ldots, 2^{n-2}p.$$

The sum of these is

$$\begin{aligned}
(2^n - 1) &+ p + 2p + \cdots + 2^{n-2}p \\
&= p + p(1 + 2 + \cdots + 2^{n-2}) \\
&= p[1 + 1 + 2 + \cdots + 2^{n-2}] \\
&= p[1 + 2^{n-1} - 1] \\
&= 2^{n-1}(2^n - 1) \quad \text{since } p = 2^n - 1. \qquad \text{Q.E.D.}
\end{aligned}$$

I should note at this stage that Euclid, although he knew a lot of maths, didn't know when $2^n - 1$ was a prime. Neither do we, exactly, but we have learned a lot more since the 17th century, when Mersenne drew attention to primes of this form. I'll say more about this at the end of this section. While I'm making a detour, it should be pointed out that there are other anachronisms in the above. For example, Euclid didn't know about algebra, but I have set out his work as if he did.

But back to perfect numbers. Euclid's Theorem raises some questions. For example, are all perfect numbers even, like the ones given by Euclid's formula $2^{n-1}(2^n - 1)$? Are other formulae possible for perfect numbers, and are there infinitely many perfect numbers?

Well, it turns out that even perfect numbers always end in 6's or 8's. In fact the first few examples alternate with 6 and 8. That made a good conjecture until the fifth and sixth perfect numbers both ended in 6. However, in 1588, Pietro Cataldi, an Italian (see Wikipedia, Pietro Cataldi), proved that Euclid's formula in the definition above, always ends in either 6 or 8.

But why should perfect numbers always be even? This turns out to be a very difficult question to settle. We do know that all even perfect numbers are of Euclid's form, thanks to a theorem proved by Euler in the 18th century (see Wikipedia, Leonhard Euler). But no one has ever found an odd perfect number, despite a lot of searching. You'll understand that remark even better if you know that there is no odd perfect number less than 10^{1500}. For this and other conditions that odd perfect numbers need to satisfy, see Wikipedia, Perfect number.

Mersenne

I want at least to mention Mersenne (see MacTutor, Mersenne and Wikipedia, Mersenne). He was a 17th-century French friar who noted that there were primes of the form $2^p - 1$, where p is a prime. Naturally these are now called Mersenne primes. The small Mersenne primes include, 3 ($p = 2$), 7 ($p = 3$), and 31 ($p = 5$). It is not too hard to show that if $2^n - 1$ is prime, then n must be prime. But it is not clear *which* primes p make $2^p - 1$ prime. It is not even known whether $2^p - 1$ is prime for infinitely many p. You might like to look up or discover more about Mersenne primes in the questions below. These primes are clearly associated with perfect numbers because Mersenne primes give perfect numbers by Euclid's Theorem.

Since it is not clear which primes p give prime $2^p - 1$, Mersenne primes are also things that mathematicians have played with. It is particularly fun for those with computing skills. To see this, in 2018, the Mersenne prime

$$2^{82,589,933} - 1,$$

was found by the Great Internet Mersenne Prime Search (GIMPS). See

https://www.mersenne.org/primes/

What use can there be for a prime with 24,862,048 digits? If not fun, well, money! There are now substantial prizes for obtaining a new largest such prime. But they also have important value for banks and other groups who want to send securely coded messages on a daily basis. More will be said about this later in Part 4. In the meantime you might like to look at Wikipedia, Largest known prime number, and ogle its size. You might also notice that the primes being found are not just Mersenne primes, though many are. By the time you get to read this there may well be new primes added to the list.

An Aside: Sum of the Numbers from 1 to n

A curious aspect of Euclid's formula $2^{n-1}(2^n - 1)$ for perfect numbers is that, by letting $k = 2^n - 1$, it takes the form $k(k + 1)/2$ – which is the sum of the numbers $1, 2, 3, \ldots, k$.

Now Euclid must have known what the sum was and I'm so sure of myself here that I'll show two ways to find the sum of the first n numbers. After all if it's good for Pythagoras' theorem to have 370 plus proofs then it's good enough for the sum of the first n numbers to have two proofs. It might just be worth looking at what I'll get for $n = 5$ first though.

$$1 + 2 + 3 + 4 + 5 = 15.$$

But what's the formula here? I'm just going to plough on. I know two ways to make progress on this.

Method 1: Let $S_n = 1 + 2 + 3 + \cdots + (n - 2) + (n - 1) + n$ be the sum we want. Now let's turn things around. Literally. Writing the sum backwards, we get

$$S_n = n + (n - 1) + (n - 2) + \cdots + 3 + 2 + 1.$$

Having written S_n forwards and backwards, I'll add the numbers that lie one above the other. So

$$
\begin{aligned}
2S_n =1 \quad &+2 \quad\quad +3 \quad\quad\quad\quad \cdots + (n-2) + (n-1) + n \\
+n \quad &+ (n-1) + (n-2) \quad \cdots +3 \quad\quad +2 \quad\quad +1 \\
=[n+1] &+ [n+1] \; + [n+1] \quad \cdots + [n+1] \; + [n+1] \; + [n+1].
\end{aligned}
$$

Each square bracket is the sum of the two numbers above it, which is $n + 1$. And the number of square brackets is n. That gives us

$$2S_n = (n + 1)n.$$

So S_n really is $(n + 1)n/2$.

The numbers $1, 3, 6, 10, 15, \ldots, (n + 1)n/2, \ldots$ are called the **triangular numbers**, because a triangular number of dots can be arranged in a triangle, like those in Figure 3.3. Surely Pythagoras, or one of his group, must have seen how to find the sum of the first n numbers.

If you read some mathematical anecdotes you may get the impression that Gauss (see MacTutor, Gauss), was the first person to prove that the sum of the first n numbers was $n(n + 1)/2$. But that has to be a lot of rot. My first piece of evidence is in Figure 3.3 where I've shown how Pythagoreans viewed triangular numbers as triangles – at least that seems

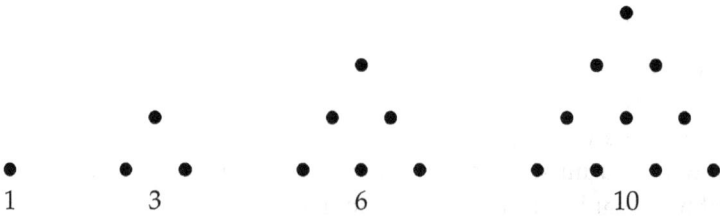

Fig. 3.3   Numbers arranged in triangle form

to be the opinion of scholars (see Heath (1981), p. 76). In other words, 6 was written as one dot on top of two dots on top of three dots. And Archimedes, not long after Euclid, managed to sum the first n squares. Then there is Euclid himself, who was able to find and prove the sum of the first n powers of 2. He even used it to produce perfect numbers! Anyone who could do that surely had to know that $1 + 2 + 3 + \cdots + n = n(n+1)/2$.

Method 2: The proof can be done using geometry, simply by combining two triangles to make a rectangle. To do this we reshape the triangles in Figure 3.3 as the right-angled triangles shown in Figure 3.4.

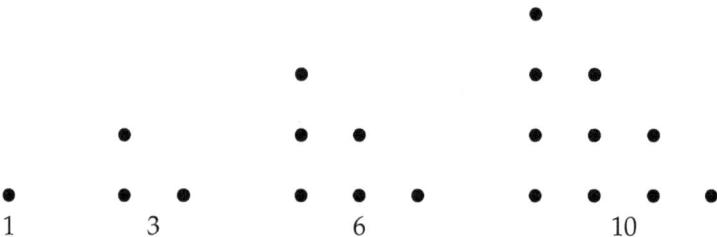

Fig. 3.4   Triangular numbers as right-angled triangles

The nth triangle now is n dots wide and n dots high. So, if we take two copies of the nth triangle and sit one (rotated) beside the other, as shown in Figure 3.5 for the case $n = 4$, we get a rectangle that is n dots high and $n + 1$ dots wide.

Counting the dots in the $(n + 1)$ by n rectangle gives $(n + 1)n$. So the number of dots in the triangle is half that, $(n + 1)n/2$.

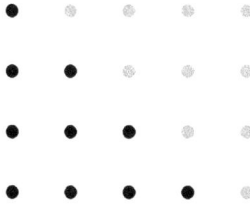

Fig. 3.5   A triangular number, twice

3.7   Pythagorean Triples

There is more to the triples than what we saw in Section 2.6. What I want to do now is to find a way that will give you all possible Pythagorean triples. Can I use $a^2 + b^2 = c^2$ to give all possible Pythagorean triples? Thus we are viewing $a^2 + b^2 = c^2$ as a **Diophantine** equation – one for which we want only integer solutions.

Can I solve that equation? Well, Euclid did. But one more thing. If you can find one Pythagorean triple you can find an infinite number. For example $(3, 4, 5)$ is a triple, but so are $(6, 8, 10)$, $(9, 12, 15)$ and $(3k, 4k, 5k)$ for any whole number k. To make life simpler, call a triple **primitive** when there is no common divisor except 1 for a, b and c. If I can find the primitive ones, I can get them all by just multiplying by k.

I have $a^2 + b^2 = c^2$, so I also know that $b^2 = c^2 - a^2 = (c + a)(c - a)$ and let $m/n = (c + a)/b = b/(c - a)$. But not just any m/n. I'll assume that m and n have no common divisor except 1 (such m and n are said to be **co-prime** or **relatively prime**). So I now have

$$\frac{m}{n} = \frac{c + a}{b} \quad \text{and} \quad \frac{n}{m} = \frac{c - a}{b}.$$

In other words

$$\frac{c}{b} + \frac{a}{b} = \frac{m}{n} \quad \text{and} \quad \frac{c}{b} - \frac{a}{b} = \frac{n}{m}.$$

Adding and subtracting these two equations and tidying up I get

$$\frac{c}{b} = \frac{m^2 + n^2}{2mn} \quad \text{and} \quad \frac{a}{b} = \frac{m^2 - n^2}{2mn}.$$

It now looks like $a = m^2 - n^2$, $b = 2mn$, and $c = m^2 + n^2$, but to conclude that I have to check that $m^2 + n^2$ and $2mn$ are co-prime, like c and b are (and similarly for $m^2 - n^2$ and $2mn$).

This follows because m and n are co-prime, but it involves some tedious checking of odd and even cases, which I'll skip.

I am finally able to conclude that if I have a primitive Pythagorean triple (a, b, c), then

$$a = m^2 - n^2, \quad b = 2mn, \quad \text{and} \quad c = m^2 + n^2$$

for some relatively prime m and n. This says that if we have a triple, then a, b and c can be written in terms of m and n. Is that an improvement? Yes, because we get all primitive Pythagorean triples by this formula, and *only* primitive Pythagorean triples. Indeed we can put any relatively prime numbers m and n into the formula and get a Pythagorean triple, because

$$(m^2 - n^2)^2 + (2mn)^2 = m^4 - 2m^2n^2 + n^4 + 4m^2n^2$$
$$= m^4 + 2m^2n^2 + n^4$$
$$= (m^2 + n^2)^2.$$

And, as I've just said, we can check that relatively prime m and n give relatively prime $m^2 - n^2$, $2mn$ and $m^2 + n^2$. For example, if we put $m = 14$ and $n = 3$, we get the triple $(187, 84, 205)$.

Consequently we get the following result.

Characterisation of primitive Pythagorean Triples: (a, b, c) is a primitive Pythagorean triple if and only if there exist relatively prime m and n such that

$$a = m^2 - n^2, \quad b = 2mn, \quad \text{and} \quad c = m^2 + n^2.$$

Such theorems are called characterisations, because they express a given property, here "(a, b, c) is a primitive Pythagorean triple", by means of an equivalent, in this case the property that "there exist relatively prime m and n such that $a = m^2 - n^2$, $b = 2mn$, and $c = m^2 + n^2$".

You may remember Pythagoras' theorem (together with its converse) says that "A triangle with sides a, b and c is right-angled if and only if $a^2 + b^2 = c^2$". This result characterises right-angled triangles by the sum of squares rule. You don't have to get out a protractor and check that there is a right angle there, only to check the equation. It is thought that in ancient times a portable (3,4,5) triangle, created by 12 equally-spaced knots in a loop of rope, may have been used by tradesmen to create a right angle by stretching the rope (Figure 3.6). There are carpentry books and YouTube videos that actually suggest this method. And you will notice that the book *The Pythagorean Theorem* by Scott Loomis (1928) was published by a Masonic Lodge, no doubt because masons are keen on right angles.

Before I move on it's worth noting the following properties of primitive Pythagorean triples.

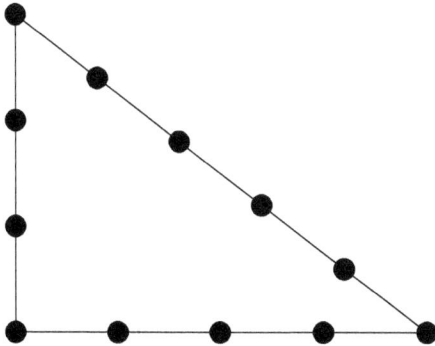

Fig. 3.6   Right angle by rope stretching

(1) One of a, b is even and the other is odd;
(2) Side c is odd;
(3) One and only one of a and b is divisible by 3;
(4) One and only one of a and b is divisible by 4;
(5) One and only one of a, b and c is divisible by 5;

That makes me think that in some way all primitive triples are just $(3, 4, 5)$.

PART 2
Liu Hui

Chapter 4

Life and Times of Liu Hui

Fig. 4.1   Stock image of Liu Hui (Wikipedia)

4.1   Introduction

The mathematician Liu Hui (Figure 4.1) is considered by some as one of the greatest mathematicians of all time. This may or may not be right, as several other mathematicians have been recommended for this crown. Rightly or wrongly though, it is certain that he was pretty good.

Like Euclid before him, and others to come in the following chapters, there is considerable difficulty in determining his birth and death dates.

MacTutor gives "about 220" and "about 280", while Wikipedia has "225 to 295". At least we know that he was alive and well in the third century CE.[1]

To give you some idea of what you can read in this chapter and the next, I want to point out four sections. Section 4.2 gives a historical background to Liu Hui's life. This contains a few notes on what happened in the third century or before it.

Section 4.3 discusses the work of some famous mathematicians between the time of Euclid and the time of Liu Hui. Section 4.4 provides a short biography of Liu. It is necessarily short because history has left no note of who his friends were, whether he played internationals for China, or what he liked to do on his days off. Again this is not unique among the mathematicians I have chosen to write about here. Section 4.5 outlines the achievements of Liu.

Chapter 5 is where the main mathematics of this Part can be found. It is based on Liu's commentary on the book *Nine Chapters on the Mathematical Art*. This is an important book, especially in China. Its aim was to provide methods that professionals, not necessarily mathematical professionals, could use to help them undertake the maths they needed in their work. It was certainly successful in this, in that it was used at university level for many hundreds of years. However, there were sections in Liu's commentary that broke new ground in mathematics. For example, independent of Archimedes he was able to produce two methods that give a value of π. We elaborate on this as well as the Chinese proof of Pythagoras' theorem, called the Gougu Rule, that may have been discovered many years before Pythagoras. Other things include methods to find square roots.

Following on from this, Section 5.6 is on Liu's book, the *Sea Island Mathematical Manual*. Liu initially wrote this as part of the *Nine Chapters*,[2] but it was separated later. This is essentially a book for surveyors and has many interesting questions on finding heights and depths of objects.

---

[1]Note that most years I use in this chapter and beyond are in the CE division of time. We are no longer in the BCE, where Euclid lived. For the rest of this book, any year is a CE year unless I say otherwise.

[2]I'll reduce the title *Nine Chapters of Mathematical Art* to *Nine Chapters* from now on. In the same way, I'll refer to *Sea Island Mathematical Manual* as *Sea Island*.

4.2   History

Aspects of the Roman Period

The Roman Republic, founded in the sixth century BCE, was an agglomeration of city states ruled by magistrates who were elected annually. The Roman Empire brought these city states under one leader as it took over much of the area around the Mediterranean Sea. The Empire lasted from 27 BCE until 1453.

Third Century BCE. During the third century, the Roman Empire almost fell apart. At that time Roman influence covered all of the regions abutting the Mediterranean Sea, along with Portugal and Britain, and extended to some areas adjacent to the Dead Sea. What caused the rapid decline were both internal and external problems. First several Roman generals vied for the position of Emperor and this led to civil wars. These wars motivated a number of attempted incursions by tribes on the borders of the Roman Empire. In addition there was the matter of climate change that lasted for several hundred years. This was called the Roman Warm Period (see Wikipedia, Roman Warm Period). It is not clear whether this was a local or global phenomenon. Anyway the weather enabled grapes to be grown in England, while some fruits in the rest of Europe wouldn't grow. Agriculture was made worse because many farmers were forced to be soldiers.

   The combination of war and weather led to a decline in the economy. It was not until the emperor Aurelian managed to defeat the Visigoths and other enemies that Rome recovered somewhat (see Wikipedia, Crisis of the Third Century).

Roman Britain. This era in Britain seems to have been a mixture of wars, local skirmishes and guerrilla warfare from the moment that Julius Caesar led his troops in the first attacks in 55 BCE, to the time they withdrew (Wikipedia, Roman Britain). One of the important local tribal leaders was Boudica (Boadicea), a queen of the Iceni tribe, who laid waste to Colchester, London, and St Albans before being defeated. She is commemorated by a statue on the Thames near Big Ben.

   The final withdrawal from Britain was totally unplanned. The problem was that Rome was under attack in Gaul (France). Honorius, the Roman Emperor of the time, essentially told generals in Britain that it was every Roman man and sympathisers for themselves and washed his hands of them. But during the Roman presence in Britain a great deal of Romanisation in the form of palaces, forums, basilicas and amphitheatres was undertaken. Much of this disintegrated after the Romans left Britain, in conflicts between local tribes.

Calendars. Caesar was not a mathematician, but he was the first Emperor and had gained that position because of his victories in war. But I don't want to talk about these. What he did accomplish was a great stride forward in the annual calendar. I tend to think of great generals of the period as dictators who ran everything by themselves. It was clear there was something wrong with the way the months moved around under the old Roman calendar and Caesar was trying to achieve some consistency. As a result, in 46 BCE, he called together a group of top philosophers and mathematicians to solve the problem. The result in 45 BCE was named the **Julian calendar**. No, Caesar was not a mathematician (and as it will appear, not a very good administrator on this occasion either), but he gets his name here because of the work that he set into motion.

So what was the problem? The Roman calendar of Caesar's time was based on 12 months that covered 355 days. Certainly by that stage everyone knew that the solar year, the time for the Earth to go round the Sun once (or maybe the Sun around the Earth depending on your viewpoint), was 365 days. Now to make the year's length work, they added in 22 or 23 days after the first 23 days of February. Then February's original last seven days were added with March happily following.

You'll notice that that means the year was 377 or 378 days long. So to fix this, they had a system whereby the 355-day year was put in between alternating years of 377 and 378 days. But then they noticed this wasn't quite right as the Sun and the calendar got out of synchronisation. As a result, in one group of 8 years every 24 years, the alternating years were only 377 days long. The net effect was to give an average year over the 24-year cycle above, of exactly 365 1/4 days. Not a very easy thing to manage or remember or even to understand, but something that would work fairly well with only a small error because of the real length of the year.

The actual difficulty of this calendar was not the complicated system, but the people who were in charge of it. They were mainly politicians who were elected every year. But there were candidates who were their mates and candidates who weren't. Consequently the extra days were likely to be added to the next year when a mate was up for election and extra days were forgotten on other occasions. As a result, the average year length over 24 years could be pretty well anything.

Note that this was a political and not a mathematical screwup that meant the calendar caused the Sun to wander all over the (calendar) place. The original purpose of the Roman calendar had been usurped. And what was the solution? Simplification and removing politicians from the system. The latter was done by making the year a fixed length of 365 1/4 days. To do this the Julian calendar added two extra months between November and December and the number of days in the various months were fiddled to make 365. A leap year was established that added an extra day to the year every four years. What's more the calendar year was deemed to begin on January 1st.

The Julian calendar spread around the Roman Empire and from there to the Mediterranean and most of the Western World until 1582. At that stage it was clear that the Julian calendar wasn't perfect, so Pope Gregory XIII brought in his astronomers and mathematicians. They made a minor tweak based on the knowledge that the solar year was 365.2425 days. This meant that every year divisible by 4 is a leap year unless it is divisible by 400. From then on, we have had the **Gregorian calendar**. *But*, it too has a problem, because the solar year is not exactly 365.2425 days. So every now and a long then, minor differences have to be made so that the Sun satisfactorily was consistent as possible with the calendar.

Now the Chinese calendar was, like many other calendars, lunisolar – it also acknowledged the moon. The earliest record of a lunisolar calendar in China is under the Zhou Dynasty in 1050 BCE. The main feature of such calendars is that they use both observations of the Sun's year and the Moon's phases. The Moon is mainly the determiner of festivals. In addition, the years are considered in 60-year cycles and that is where the animals come in. They exist because the Chinese zodiac has 12 animals (Rat, Ox, Tiger, Rabbit, Dragon, Snake, Horse, Sheep, Monkey, Rooster, Dog and Boar), which repeat five times each in the 60-year cycle.

4.3 Famous Mathematicians

Archimedes

Archimedes lived roughly around 287–212 BCE. He is another person who is said to have been the best mathematician in the world; he was certainly one of the best scientists in the ancient world. And one of his discoveries is perhaps the most entertaining story in the world of mathematics. Basically, his king ordered a golden crown and when he got it, he had reason to believe that it wasn't all gold. So, the king turned to Archimedes for help in determining the quality of the crown. Of course, Archimedes put a great deal of effort into the solving thereof, but it wasn't until one day he was having a bath that the whole thing clicked into place. He noticed that when he stepped into the bath, water flowed out. And that was how the solution presented itself to him. He was so excited about having solved the king's problem that he ran out into the street naked and yelled out "Eureka!" (meaning "I have found it!").

Ever since, perhaps not quite, a Eureka moment is when you get a sudden idea and whatever you are working on becomes clear. If you are a student, it can happen when you are doing your homework. I had one in an exam once, when the topic that I had struggled with for a year suddenly made sense. Poincaré had one getting on a bus and suddenly he could solve a big problem he was working on (MacTutor, Poincaré). And Hamilton finally found quaternions (Wikipedia, Quaternions) while walking along a canal in Dublin. He was so excited he got out his knife and carved their equations on Broome Bridge, where there is a memorial plaque today.

Eureka moments are well documented (Wikipedia, Eureka). They generally come after a great deal of work, and you are doing something else other than looking at the thing you are tussling with. And they really do come to anybody.

Unfortunately, the general belief among experts is that Archimedes didn't run naked down the street saying "I have found it" in Greek. But it makes for a good story, and he *had* solved the problem. It had led him to the **Archimedes principle**:

> Every body in a fluid, whether submerged or floating, experiences an upward force. This force is equal to the weight of the fluid displaced.

What this means for Archimedes' bath is that if the water was level with the top of the bath when he got in, the weight of the water that spilled out of the bath was equal to the force upwards on his body.

I'm just wondering how long it took him to come up with this principle and what was the Eureka moment he had? First of all you would have thought that a great deal of experimenting and data gathering would be needed before you could come up with this principle. Was the excitement of Archimedes caused by this work all coming together in a formal description of the principle? Or was all the yelling caused by the fact that he had been wanting a practical application of his principle and he was happy because he saw how to apply it to the crown?

At some time anyway, he applied the one to the other. But how? Well, there was a crown that might or might not have been all gold. Suppose the crown maker had dropped in a bit of silver. Silver is lighter than gold, density wise. So if the crown hadn't been the same weight as the gold that the king had passed over, the king could have easily determined the presence of a fiddle.

That meant that the crown was the same weight as the given gold. But that would have meant that it wasn't the same size as a gold one. It had to be bigger. Putting it into water then, would have caused more water to be pushed out of the bath than would have been pushed by gold alone. And by the new principle there would have been more push up on the scam crown. How to tie this all together? Was it too hard to accurately measure the volume displaced by the two objects? Or did Archimedes have a set of scales? Put the scales in the bath and add the gold to one side and the crown to the other. If the volume of the crown was bigger, then more water was pushed up than in the gold case, *and* that meant a greater push of buoyancy on the crown. So if the crown side of the scales went up, the crown manufacturer needed to run, naked or not!

Being a great mathematician, Archimedes had more things on the go than playing with crowns. One of his other major achievements was to find an approximation to π, more accurate than 22/7, the one you possibly learned in your schooldays. In fact, Archimedes showed that

$$3\frac{10}{71} < \pi < 3\frac{1}{7}.$$

Now it's very hard to measure either the circumference or the area of a circle directly. For the former it's hard to get a bit of string to keep to the curve, so even if you know $C = 2\pi r$, it's hard to find π by the circumference route. And even if you know that the area of a circle is given by $A = \pi r^2$, breaking the area up into very small squares is still not the easiest thing to do, especially when there are little non-square areas near the circumference.

But let's say that Archimedes put a square in a circle so that the circle passed through the corners of the square. We can construct this square because first we know the centre of the circle and we can draw a diameter that can be one diagonal of the square. We have a construction that will put a perpendicular to this diameter thanks to Euclid, and so we can find the other diagonal of the square. Then, by constructing perpendiculars to the diagonals of the inner square, we can make a square outside the circle, touching it at the corners of the inner square (left picture in Figure 4.2).

Next, drawing diagonals of the outer square, we can make octagons inside and outside the circle (right picture in Figure 4.2).

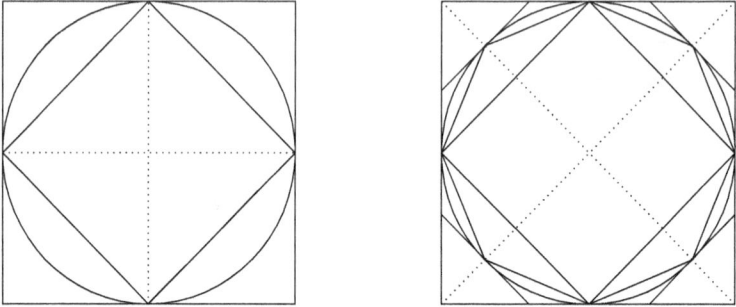

Fig. 4.2   Constructing polygons, inside the circle and outside

We can continue to double the number of sides of the polygons inside and outside the circle by bisecting angles. We can also start with a different constructible polygon, such as an equilateral triangle. And all the lengths involved can be calculated by Pythagoras' theorem. Using this method for a polygon of 96 sides, Archimedes found that π is between 223/71 and 22/7, as mentioned above.

One of his other achievements was inventing a screw device to bring up water from a river or other water source (Figure 4.3). It was used extensively in a large number of places because of its simplicity. There are companies that produce them, even today. It is still wonderful that water can follow gravity *down* and yet move *up* and out of the device.

Fig. 4.3   Archimedean screw, from *Chambers Twentieth Century Dictionary*, 1908

Archimedes was also famous for his very effective war machines. These were supposedly invented to stave off attacks by Romans on the city of Syracuse. I'm going to look at three of these. First, his Claw. This was like a big crane with a hook on the end. The idea was essentially fishing for ships. The hook was lowered until it caught the ship and then shaken to destroy it. This sounds more like a theoretical idea than a practical one.

In Figure 4.4 (from Wikimedia), I've shown a possible "Heat Ray". It consists of several mirrors, ideally arranged along a parabola, to reflect the light of the Sun onto a ship. The heat generated sets the ship on fire. The diagram shows the Sun absurdly close to the mirrors, of course. In reality the rays from the Sun are virtually parallel, because the Sun is so far away, and if they are also parallel to the axis of symmetry of the parabola they reflect to a single point, called the **focus**. Evidently Archimedes was aware of this property. Incidentally, the term "focus" was introduced by Kepler, from the Latin word for fireplace.

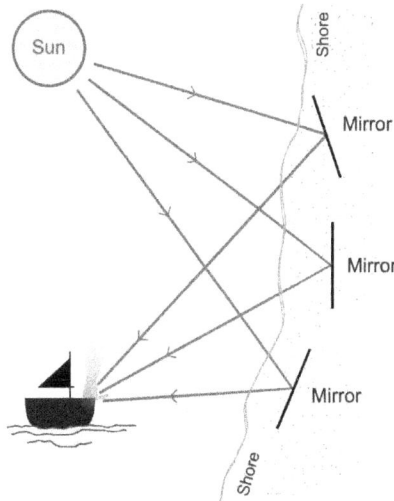

Fig. 4.4   Archimedes' Heat Ray (from Wikipedia, Archimedes)

The same idea, on a gigantic scale, is used today in certain solar power stations (see Wikipedia, Concentrated solar power). One of them, with about 10,000 mirrors, is shown in Figure 4.5.

In addition, Archimedes developed a catapult powerful enough to throw heavy objects towards an enemy.

Fig. 4.5   Concentrated solar power (from Wikipedia, Concentrated solar power)

Diophantus

Diophantus' life in Alexandria overlapped Liu Hui's as the former lived from roughly 200 to 284. There is very little that we know about his life, but that seems to be par for the course for people, even important people, who lived about 2,000 years ago. However, there is a problem about him that appeared in a Greek Anthology about 500. It purports to give his age. It goes like this:

> ...his boyhood lasted 1/6th of his life; his beard grew after 1/12th more; he married after 1/7 more; and his son was born 5 years later; the son lived to half of his father's age, and the father died 4 years after the son.

So now you know for how many years Diophantus lived.

Let me just interpose Hypatia, who was briefly mentioned in Section 1.5. She, let me underline that, *she* may not be the first female mathematician that we know of. Before Hypatia was a person called Pandrosion, who may have been female, but we know very little about her. (See Wikipedia, Pandrosion.) Hypatia lived from about 360 to 415. One thing that we know about her is that she wrote a commentary on Diophantus' *Arithmetica*.

It's just worth mentioning that when we say "commentary" in this historical context, it was more than just a review. Generally it was totally based on what the original author said, though the commentary might omit certain things. On the other hand, it would contain extensions or even additions by the commenting author. In Section 5.1 we will find that Liu was into making commentaries too.

Diophantus wrote 13 books called the *Arithmetica*, but we only know something about six of these. It seems that Hypatia wrote a commentary on just the first six books and the remaining books didn't get the publicity that they deserved. Nevertheless, the seventh one may have survived through Arabia. An Arabic manuscript that claimed to be Books IV to VII was written before 912, and was found in 1968. But there is a style difference between them and the *Arithmetica*, which suggests that the manuscript wasn't a direct translation of Diophantus' work. It seems that the book may have been a commentary of his work and it may have been written by Hypatia.

Anyway, we know that the *Arithmetica* contained 130 problems. The mathematics of the problems is so original that the equations involved are now called Diophantine equations and the solution methods are called Diophantine analysis. The equations are polynomial equations, usually in two or more variables, and what makes them "Diophantine" (as we mentioned in the case of $a^2 + b^2 = c^2$ in Section 3.7) is that rational or integer solutions are sought. It is also worth noting that Diophantus used a special notation, called "syncopated algebra", that has the smell of algebra as we know it – it has some symbolism but not all the characteristics of the symbolic algebra we currently use. For instance, he wrote

$$\mathrm{K}^\upsilon \overline{\alpha}\ \zeta \overline{\iota}\ \pitchfork\ \Delta^\upsilon \overline{\beta}\ \mathrm{M}\overline{\alpha}\ \overline{\iota}\sigma\ \mathrm{M}\overline{\varepsilon}$$

where we would write $x^3 - 2x^2 + 10x - 1 = 5$. (For a way to translate Diophantus's algebra, see Wikipedia, Arithmetica.) As these "syncopated" expressions are somewhat like algebraic notation, when you know how to read them, Diophantus has been called the "Father of Algebra". (But he is not the only one.)

Diophantus worked on linear, quadratic, and cubic equations, but he was usually interested only in solutions that are rational. In just one case, the equation $y^3 = x^2 + 2$, does he mention only the integer solution, $x = 5$, $y = 3$. This example became the inspiration for many later mathematicians, as we will see. Basically, he ignores negative numbers (such as the solution $x = -5$ of $y^3 = x^2 + 2$) as not giving valid answers to real questions that might involve objects such as books. In his day seeking rational solutions was a big step forward from, though actually easier than, seeking whole number solutions. (For example, the equation $y^3 = x^2 + 2$ has infinitely many rational solutions.) The reason is that his methods for finding rational solutions allow *geometry* to be used in solving equations.

I'll illustrate the geometry behind his methods with an example. Book VI, Problem 18 of the *Arithmetica* finds a rational solution of the equation we would write as

$$y^2 = x^3 - 3x^2 + 3x + 1.$$

Now this equation has an "obvious" solution $x = 0$, $y = 1$, but that is too easy for Diophantus. Instead, he substitutes $y = 3x/2 + 1$ and gets

$$x^3 - 3x^2 + 3x + 1 = \left(\frac{3}{2}x + 1\right)^2 = \frac{9}{4}x^2 + 3x + 1.$$

This simplifies, by cancelling $3x + 1$ from both sides, to

$$x^3 = \frac{21}{4}x^2,$$

which (cancelling x^2 from both sides) gives the rational solution $x = 21/4$. Then, since $y = 3x/2 + 1$, y is also rational, in fact $y = 71/8$.

Thus Diophantus has found a far from obvious rational solution, thanks to the miraculous substitution $y = 3x/2 + 1$. How did he know this would work? Well, he could, and probably did, foresee that this substitution would create the $3x + 1$ that cancels from both sides. However, there is a more interesting geometric interpretation, shown in Figure 4.6.

If we interpret $y^2 = x^3 - 3x^2 + 3x + 1$ as the equation of a curve, as was first done in the 17th century, it will be noticed that $y = 3x/2 + 1$ is the equation of the *tangent* to this curve at the point $(0, 1)$. Thus the tangent leads from the point representing the "obvious" solution to a point representing a non-obvious solution. In the 17th century, Fermat and Newton realised that the tangent at any rational point on a cubic curve would lead to another rational point, which led to the method becoming known as the "Diophantus tangent" method. This realisation led to a fusion of number theory, algebra, and geometry that is typical today – and I like to think it would have delighted Euclid, and Diophantus.[3]

---

[3]Incidentally, the tangent method on the curve $y^3 = x^2 + 2$, starting at the point $(5,3)$, gives $(383/1000, 129/100)$, the first of infinitely many rational solutions.

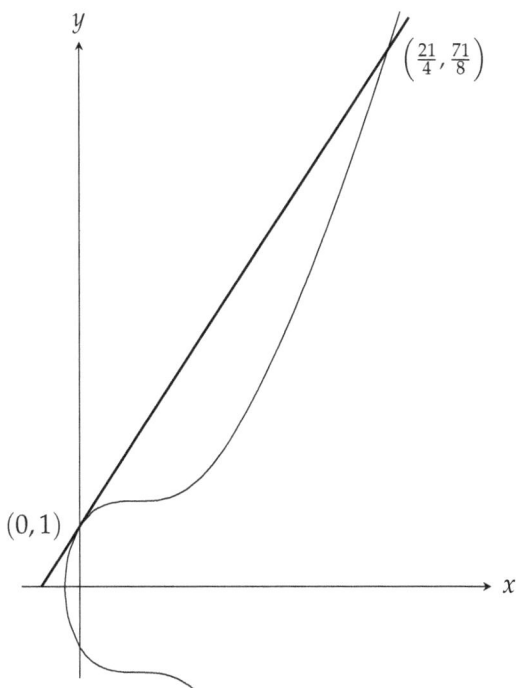

Fig. 4.6　Cubic curve $y^2 = x^3 - 3x^2 + 3x + 1$ and tangent

Just a footnote here. Claude Gaspar Bachet Sieur de Méziriac in 1621 translated the *Arithmetica* into Latin. It was one of these books that Fermat was reading when he scribbled in the margin, and the long life of Fermat's last theorem began (see Chapter 13).

Diophantus was interested in what numbers were sums of squares too. For instance, he seemed to believe that

numbers of the form $4n + 3$ cannot be the sum of two squares;
numbers of the form $24n + 7$ cannot be the sum of three squares; and
every number can be written as the sum of four squares.

Did Diophantus have a proof for these statements? It is just possible that such results were in one of the lost books of the *Arithmetica*, though writing proofs was not Diophantus' style. He may however have *understood* why the first two are true. For the first it suffices to consider the possible sums when the numbers being squared are odd or even. It turns out that the sum must be of the form $4n$, $4n + 1$, or $4n + 2$. (If you need help with this, see Section 13.8.) But a proof of the third result would have been dramatic. Because of Bachet's translation of the *Arithmetica*, the third result became known as Bachet's Conjecture. It became Lagrange's "four-square theorem" when proved in 1770. In 1797 or 1798 Legendre was also able to prove that a number is the sum of three squares if and only if it is not of the form $4^m(8n+7)$ for positive integers m and n.

The Ten Chinese Classical Maths Texts

In 581, after about 300 years of an unsettled and divided China, the Emperor Wen of Sui, came to power. The Sui dynasty only lasted until 619 when Wen's son lost control after a fairly vicious reign. However, thanks to Wen, a number of positive decisions were made and these were continued by the succeeding Tang dynasty.

One of these initiatives include the Kaihung legal code that simplified earlier laws. Simultaneously an effort was made to ensure that local officials knew what the laws were and actually enforced them. (You'd think that those two things were reasonably important.) The Kaihung system developed into the later, and better known, Tang Code (see Wikipedia, Tang Code). This code became the basis for legal systems elsewhere in East Asia.

Further, the administration of the country was revived to the system of three departments and six ministries that had begun during the Han dynasty. This system was accepted by pretty well every future dynasty.

On the education side, imperial exams were established for the civil service. As a result, future civil servants thought of their job as a privilege rather than a duty. By the mid-Tang dynasty, these exams were the main way for people to gain office.

Education was extended further by the Tang dynasty when the astronomical observer told the emperor that many of the classical mathematical books used by the Imperial Academy contained errors. As a result, in the middle of the seventh century, Li Chunfeng led a group of top academics to review the books. These were subsequently used at the National University. From 1084, these books, were called the Ten Mathematical Classics or the Ten Computational Canons.

If you are a perceptive reader, you will note that twelve books are listed below, not ten, but these are the books given in MacTutor. However, in some listings *Shushu Jiyi* and *Sandeng Shu* are omitted. I'm not sure what is going on here, but in Chemla (2012) *The History of Mathematical Proof in Ancient Traditions* Table 15.2 on page 516 includes twelve books too. In addition, we read on p. 519 that

> The conventional identification of the twelve treatises constituting the curriculum is found in a number of modern works ...

So I'm sticking with the 12 Ten Mathematical Classics.

Table 1. Ten Mathematical Classics in approximate chronological order.

100 BCE–100	*Zhoubi Suanjing* (Zhou Shadow Gauge Manual)	
Early third century	*Shushu Jiyi* (Notes on Traditions of Arithmetic Methods)	
Third century	*Jiuzhang Suanshu* (Nine Chapters on the Mathematical Art)	
Third century	*Haidao Suanjing* (Sea Island Mathematical Manual)	
280–472	*Sunzi Suanjing* (Sun Zi's Mathematical Manual)	
425–458	*Xiahou Yang Suanjing* (Xiahou Yang's Mathematical Manual)	
430–490	*Zhang Qiujian Suanjing* (Zhang Qiujian's Mathematical Manual)	
429–500	*Zhui Shu* (Method of Interpolation)	
Fifth century	*Wucao Suanjing* (Mathematical Manual of the Five Administrative Departments)	
Seventh century	*Jigu Suanjing* (Continuation of Ancient Mathematics)	
?	*Wujing Suanshu* (Arithmetic methods in the Five Classics)	
?	*Sandeng Shu* (Art of the Three Degrees; Notation of Large Numbers)	

These books were the basis of the mathematical education of 30 students who were recruited each year from the "lower levels of society". There was a weaker class who worked on basic maths that was practical and another class that learned techniques. Seemingly putting universities of today to shame, the students studied for seven years. It does worry me a bit, though, that the upper group had an oral exam including questions that required students to complete sentences taken from the Ten Classics. That doesn't seem to promote higher level learning. However, Chemla's book (cited above) provides evidence that students were given problems in the exams of both courses, similar to problems in the books being studied. Students passed if they were successful in six of ten problems.

4.4   Biography of Liu Hui

About all that is known for sure about Liu Hui is that he lived in Cao Wei, one of the states involved in the Three Kingdoms wars (Chapter 1) and his life was lived almost completely during those wars. A large percentage of the population died during those wars, so Liu Hui was lucky not to be one of them. Liu Hui's big, and I mean big, claim to fame is that he wrote a commentary on the *Nine Chapters on the Mathematical Art* which hit the top ten mathematical classics of Section 4.3 above. In the Preface to his commentary Liu Hui suggests that he wrote not just to update the work, but also because many copies had been burnt by Qin Shi Huang, 213 BCE, in his attempt to improve education! Lui was probably mistaken about the age of the original *Nine Chapters*, but nevertheless his motives resemble those of Bourbaki, mentioned in Chapter 1. Regardless of this, Liu had a great influence on the future development of mathematics in China.

Figure 4.7, from Wikipedia, Three Kingdoms, shows the main regions of China in 263, the year when Liu Hui published his version of the *Nine Chapters*. The state where he lived is Cao Wei but, as you can see, this covers a lot of ground.

However, from his works certain things have been deduced. For instance, in MacTutor, we read

> ... we can see that he is an outstanding mathematician with a deep understanding of difficult concepts. He is also highly original, coming up with ideas which rank him among the leading mathematicians of all time.

While Shen *et al.* (1999) suggest that

> The techniques Liu employed are typical of a teacher of skill, patience and tireless zeal.

It has also been suggested that he may have held a senior administrative position because of the way he seems to care for people and their welfare, as well as the country's economy.

Fig. 4.7   The Three Kingdoms in 263 (from Wikipedia, Three Kingdoms)

It's interesting at this point to note a fundamental difference between Liu Hui's work and that of Euclid (see Part 1). As we have seen, Euclid's *Elements* contained a formal development of the mathematics known at that time. It started with definitions and axioms and theorems were logically produced from what came before. On the other hand, Liu's work was largely about practical problems that might be used in public areas such as taxation and private areas such as surveying and engineering. Solutions are given to the problems but alongside them was insight that gives a feeling for why a given algorithm might work. But not all of the work is practical. For example, Liu finds the value of π more accurately than would be needed in field work.

4.5   Liu Hui's Achievements

There is a sense in which Liu Hui only achieved one thing and that is writing the commentary to the *Nine Chapters*. Like Euclid's *Elements*, this was a book that lasted for at least a millennium and influenced the development not only of Chinese mathematics, but of mathematics throughout East Asia. This was because *Nine Chapters* was used in Japan and other countries in the region. *Nine Chapters* was a very influential book. It's worth noting that Liu was chosen to write the commentary because of his outstanding mathematical skill. This would have been a great honour for him.

It has to be said that Liu's third century commentary was a thorough review of the document. Apart from correcting errors in the extant copy, he added a theoretical basis for much of the work as well as providing new material.

The book also contains the solution of linear equations using what we would call Gaussian elimination today. However, Liu gives greater insight into the method and its underlying algorithm, which amounts to a proof that the algorithm is correct. The same can be said of his other work such as finding roots of numbers and values of π.

I haven't mentioned Liu's work on solids because, as with Euclid, in Chapter 2, some material had to be omitted because of space. However, there are parts of Chapter 5 of the *Nine Chapters* devoted to the volumes of solids such as prisms, pyramids, wedges and truncated cones. Apparently in the extant version of *Nine Chapters* from which Liu was working, there was a formula for the volume of a sphere. He knew the formula was incorrect, but admitted that he didn't know how to rectify it.

Chapter 5

Mathematics of Liu Hui

5.1 Nine Chapters on the Mathematical Art

As I have said earlier, the *Nine Chapters* is a little like Euclid's *Elements* (see Chapter 1) in that it contains too much material to discuss every section of it. As a result, I have taken a selection of topics for this chapter.

The Gougu Rule

First I need to say that Gougu isn't a person. Gou (pronounced "Go") and Gu ("goo") are the two non-hypotenuse sides of a right triangle. Figure 5.1 shows these sides and adds that the hypotenuse is Xian. (The rule is sometimes referred to as the Gougu Xian rule.) This word represents a tight chord drawn between the ends of the hypotenuse. Gou means longest of the perpendicular sides of the right-angled triangle and Gu, the shortest.

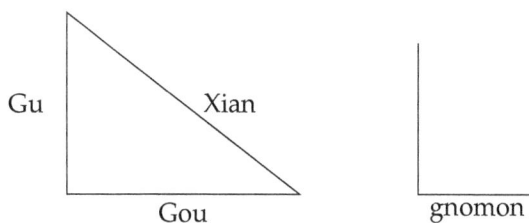

Fig. 5.1   Gou, Gu and Xian

The right-angled shape on the right in Figure 5.1 represents a thing called a **gnomon** (pronounced "no mon"). This is a right-angled tool that I think would have had a chord attached at the end of one of its sides that could be applied as a plumb line to make one side vertical and consequently, the other side horizontal. (The Greeks also used the gnomon, and indeed "gnomon" is a Greek word.)

In the *Zhoubi Suanjing* (the first book in my list of the Ten Classics) there is a nice quote, see Zhoubi Suanjing, MacTutor, that gives an idea of the many uses of the gnomon. It goes like this:

> Duke of Zhu: How great is the art of numbers? Tell me something about the application of the gnomon.
>
> Shang Gao: Level up one leg of the gnomon and use the other leg as a plumb line. When the gnomon is turned up, it can measure height; when it is turned over, it can measure depth and when it lies horizontally it can measure distance. Revolve the gnomon about its vertex and it can draw a circle; combine two gnomons and they form a square.

This exchange suggests that both arms of a gnomon were the same length, but we'll see many cases later where they're not. It's clear how this could be used for surveying, and examples are given in Section 5.6 below. Astronomers could also use it to measure lengths of shadows which was a great help for constructing reliable calendars. It is possible that Liu went to Luoyang (in the west of Henan province) for some astronomical work, because it is said that he measured the Sun's shadow there. It is possible that this was related to calendar making.

If you were thinking that this emphasis on right angles has a sort of Pythagorean feeling, right on. It turns out that the Chinese knew that

$$\text{Gou}^2 + \text{Gu}^2 = \text{Xian}^2.$$

This was proved in about 1100 BCE, 550 years before Pythagoras. The traditional Chinese proof, and possibly the first proof, is contained in Figure 5.2. This is to be found in both *Zhoubi Suanjing* and the *Nine Chapters*.

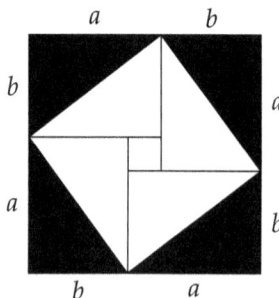

Fig. 5.2 The proof of the Gougu Rule from *Nine Chapters*

The traditional proof is wordy and hard to follow, at least for me. So first I'll give an algebraic proof – the way we would do it today – and then explain how the algebra can be simulated by geometry, giving a proof as it might originally have been discovered.

The big square has side $a + b$, where a and b are the smaller sides of the triangle, so its area is $(a + b)^2 = a^2 + 2ab + b^2$. The triangle is obviously half the rectangle with sides a and b, so its area is $ab/2$. Thus, subtracting four (black) copies of the triangle from the big square leaves

$$(a + b)^2 - 2ab = a^2 + 2ab + b^2 - 2ab = a^2 + b^2 = \text{Gou}^2 + \text{Gu}^2.$$

On the other hand, the big square minus the four black copies of the triangle is obviously the square on the hypotenuse, Xian^2. Bingo – Gougu!

Now to make this proof completely geometrical, we need only explain why $(a + b)^2 = a^2 + 2ab + b^2$. Figure 5.3 shows why. You only have to replace the algebraic terms with geometric ones: put "the square of side a" instead of a^2, "the rectangle with sides a and b" instead of ab, and so on. (This theorem, incidentally, is in Euclid's *Elements*, Book II, Proposition 4.)

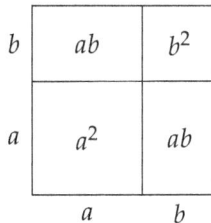

Fig. 5.3   The square on a sum of lengths

Hence the Gougu Rule follows from Figure 5.2 and Figure 5.3.

I just want to say that, when solving mathematical problems, especially word problems that link to the real world, always look for what is basically happening in the problem. Most of the time the real-life aspects can be ignored except where they provide mathematical aspects. Having turned the problem into something purely mathematical, every other aspect can be at least temporarily forgotten. The same is true for a proof. Wading through the "first small side" and so on, complicates the whole thing because of the extra words. Letting $a = $ the first small side means that we need only concentrate on the symbol a and its manipulation. We can worry about what a means at the end after the maths part has been settled. That's why algebra, when it finally came, greatly improved the mathematician's life, if not the students' lives.

Perhaps there is another aspect of maths that would have made life even easier. What changes might have been made to mathematics if the computer had come before algebra? What changes might have been made to statistics if the computer had come before algebra?

By the way, I have proved the Gougu/Pythagoras' theorem because it plays a role in what I have to say about the mathematics Liu does in the *Nine Chapters* and is basic to much of his *Sea Island* book.

5.2 Counting Boards

The Chinese invented counting boards, a method for handling basic arithmetic that dates back to the fourth century BCE. It might be called the first computer. These boards used rods or sticks as "counters". I'll show how these could be used for adding relatively small numbers. It should be easy to see how to extend a board to higher numbers. But first I have to admit that the counting boards I am showing are not necessarily the same as those the Chinese used. They probably used a vertical system on counting boards because they used a vertical rather than horizontal system of writing. For the same reason they wrote units, tens, hundreds and so on, from the right to the left.

There are two things to explain before we can do any arithmetic. These are rods and the board. Rods are relatively easy. They were made with bamboo or ivory or anything that was handy. The rods could be oriented either horizontally or vertically to represent numbers from 1 to 9 as shown in Figure 5.4. Horizontal rods better resemble Chinese characters, where 1, 2, 3 are depicted by 1, 2, and 3 horizontal strokes respectively. This is a decimal system, but with numerals that are more intuitive to read than those in the West.

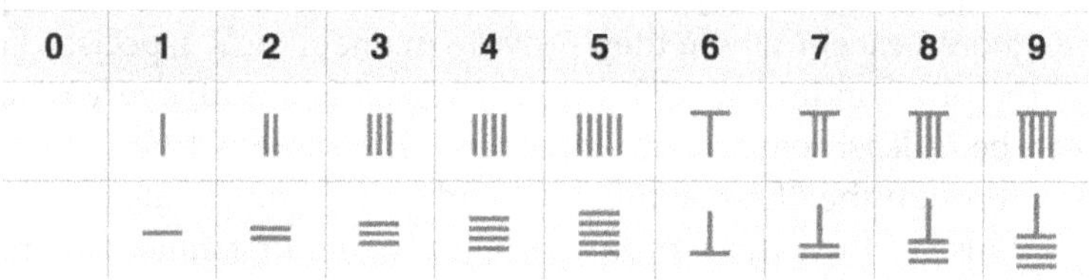

Fig. 5.4   Rods and numbers (from Wikipedia, Counting rods)

The counting boards are essentially squared "boards" that may have been made out of wood or stone. Part of a board is pictured in Figure 5.5. Boards were cheap and easy to produce, but the original ones are very rare because the cheap material decayed with time, and few are left.

Fig. 5.5   A counting board (from Wikipedia, Rod calculus)

The boards are set out so that numbers involved in calculations are entered horizontally. The squares on the board represent decimal place values. So the right-most column was the units column, the next column from the right was for tens, then came hundreds, etc. The three numbers in black rods represented in Figure 5.6 are 619, 414 and (their sum) 1033.

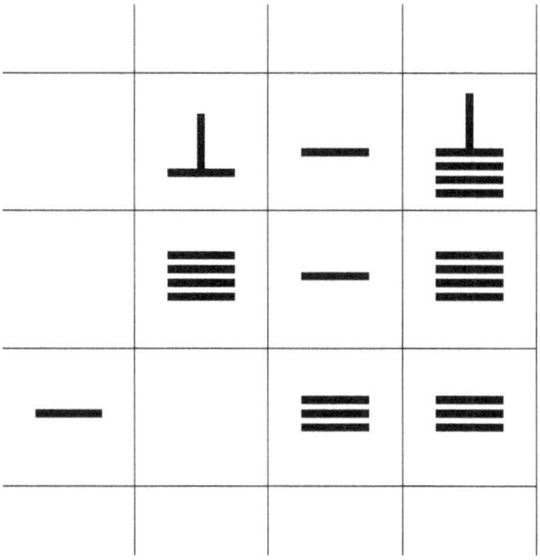

Fig. 5.6   Numbers on a counting board

At the time these boards were in use, zero symbols were not yet in use in China – but (as on the abacus) they weren't needed because a zero was simply shown by a blank square on the board. The advantage of the counting boards is that they provide a way to show place value.

Addition is relatively easy and intuitive. It's essentially the add and carry method we use today. Starting from the right in Figure 5.6, the sum of 9 and 4 rods is 13. So there we get a 3 on the right-hand end of the third column. Ten has to be carried to the second column to the right as "one", and the two black rods are then added to get 3.

In the hundreds column we have 6 + 4 which gives 10 hundreds or 1 thousand. For us there would be an entry of 0 in the hundreds column and 1 in the thousands. But there was no need for the zero on the board because a blank space told them there were no hundreds in the answer. A square on the counting board served as a placeholder, so a lack of a zero symbol was no disadvantage.

Subtracting 414 from 619 is straightforward and produces another blank entry. Multiplication and division are also possible using counting boards. (See Wikipedia, Rod calculus.)

About the 14th century, counting boards were replaced by the abacus. Again, it seems to have been a Chinese invention, though independently invented in several other civilisations. I'll discuss the role of the abacus in Europe in Chapters 7 and 8.

Negative numbers could also be catered for on counting boards. This was done by using red dots on rods for positive numbers and black for negative numbers. It's an interesting difference from our use of "in the red" to mean that we are in debt – our red numbers are negative.

5.3 Simultaneous Equations

I'll start this with a problem. The method that I have used is not the method that Liu used but the method here is an introduction to the method in the paddy problem below. This method is a general one for solving simultaneous linear equations.

The Modernised Yin-Buzu Problem

A group of friends are settling the bill at a restaurant. Everyone contributes the same amount. It turns out that if they each put in $40 they would be able to leave a tip of $15. On the other hand, if they each gave

$35, they would be short by $20. How many people were round the table and what was the cost of the meal?

We'll do this on the counting board in Table 2 below. But to help you comprehend the counting board calculations, I first set out this problem the way we would do it today. We would say something like: let $n =$ number of friends and $p =$ price of meal. Then we get

$$40n - p = 15 \qquad \text{when each friend pays \$40,}$$

and

$$35n - p = -20 \qquad \text{when each friend pays \$35.}$$

Subtracting the second equation from the first eliminates the cost of the meal, giving

$$5n = 35.$$

Finally, dividing both sides by 5 gives

$$n = 7.$$

From this we can calculate p, the price of the meal, by substituting $n = 7$ in either of the equations involving p. (And it's good to substitute in both, as a check on the answer.)

In these calculations the letters n and p are really only "placeholders" for their coefficients, which are numbers we subtract and divide. Another way to keep track is to put the coefficients of n in the first column of a table, the coefficients of p in a second column, and the right-hand side of the equation in a third column, as in Table 2.

Table 2. Calculations for equations.

Each friend's payment	Number of meals	Net payment
40	−1	15
35	−1	−20
5	0	35
1	0	7

What I have done in the third row is to subtract all of the second row entries above from the first ones. This eliminates the meal, so the third row says that if the friends pay $5 each and there is no meal to pay for, they would pay $35. The final row shows that if each person pays $1 the total payment is $7, so there must be 7 persons. Then, since 7 persons in the first instance pay 7 × $40 = $280, which is $15 more than the cost of the meal, the cost of the meal must be $265.

Look no algebra! (Or, more precisely, no symbols for unknowns.)

You should be able to see that life is much easier if we have a zero entry in one of the columns. In such cases, you get the other unknown (number of persons or price of meal) almost straight away.

And now for a second problem that uses the same method but with three variables, almost straight from the *Nine Chapters*. I've simplified it to avoid fractions. I've also written the equations horizontally, rather than the vertical way that is used in the *Nine Chapters*. To help you to see the operations on the rows, I've divided the table into three parts with three rows in each. The second of these parts is what happens after operating on rows two and three by row one, while the last part is operating on row two itself and then operating on row three with row two.

Grain Problem

Suppose that the day before yesterday you have harvested three bundles of top grade paddy, two bundles of medium grade paddy, and one bundle of low grade paddy. After all of this has been processed you have 27 kg of grain. Yesterday you got two bundles of top grade paddy, three bundles of medium grade paddy, and one bundle of low grade paddy. That yielded 25 kg of grain. Today your production was one bundle of top grade paddy, two bundles of medium grade paddy, and three bundles of low grade paddy that produced 17 kg of grain. How much grain does one bundle of each paddy grade give you? Table 3 lays out the solution.

Table 3. The grain problem.

Top grade	Medium grade	Low grade	Yield	Operation
1	2	3	17	
2	3	1	25	
3	2	1	27	
1	2	3	17	
	−1	−5	−9	(i)
	−4	−8	−24	(ii)
1	2	3	17	
	−1	−5	−9	
		12	12	(iii)

The operations (i), (ii), and (iii) that get me to the third stage are:

(i) multiply row 1 by 2 and subtract it from row 2;
(ii) multiply row 1 by 3 and subtract it from row 3;
(iii) multiply row 5 by 4 and subtract it from row 3.

(Note, incidentally, the confident use of negative numbers: recognising that subtraction of a negative is the same as adding a positive.)

Just as I wanted, the result is the three "zeros" in the bottom left-hand corner. From here life is easy. Now

(a) from the last row, I know that 1 bundle of low grade paddy yields 1 kg of grain ($1 = 12/12$);

(b) from row 8, I know that 1 bundle of medium grade paddy yields 4 kg of grain ($4 = 9 - 5$);

(c) substituting the 1 and 4 in row 7, we get high grade paddy yields 6 kg of grain ($6 = 17 - 2 \times 4 - 3 \times 1$).

Now isn't that pretty? All that I've done, thanks to Liu, is to play with some numbers – it's just arithmetic. It's less work than trying to solve the three equations

$$x + 2y + 3z = 17$$
$$2x + 3y + z = 25$$
$$3x + 2y + z = 27,$$

in the sense that you don't have to carry the baggage of symbols x, y, z. Also, it highlights the really essential actions in the solution, which are **row operations** on numbers. We do the same today when we solve linear equations using **matrices**, by reducing to "triangular form".

Liu used essentially these operations, except that he used columns instead of rows, in accordance with Chinese writing style.

The rules that can be used to simplify a system of linear equations are called the Fangcheng Rules. A number of systems of linear equations can be found in Chapter 8 of the *Nine Chapters*. Mostly there are the same number of unknowns as there are equations. However, in Problem 13 of that chapter there is a system that has less equations than there are unknowns. This is thought to be the first time that such a problem was ever posed. In this case the values of the variables cannot be found explicitly, but some of the variables can be found in terms of the other variables.

The Fangcheng Rules also appear to have been published first in the *Nine Chapters*. Nevertheless in the West, this method is known as Gaussian elimination. That was clearly a mistake, but one that wasn't made until the 1950s! For the fuller story, which involves Newton, see Wikipedia, Gaussian elimination.

5.4   Approximations of Square Roots

It is almost certain that Chinese mathematicians knew how to find square roots before the *Nine Chapters*. What Liu did was to give a mathematical proof that the method works. Let's start this section with an example. How might Liu have found the square root of 961?

Well, I'll begin with the square root of 900 because that is easier. It's not too hard to see that $30 \times 30 = 900$, so $\sqrt{900} = 30$.

Now look at Figure 5.7. First, the larger square represents an area of 961 units. I've found a part of it that has 900 units. If I could find the value of a, then I would know the dimensions of the larger square and that would give me $30 + a$ as the square root of 961.

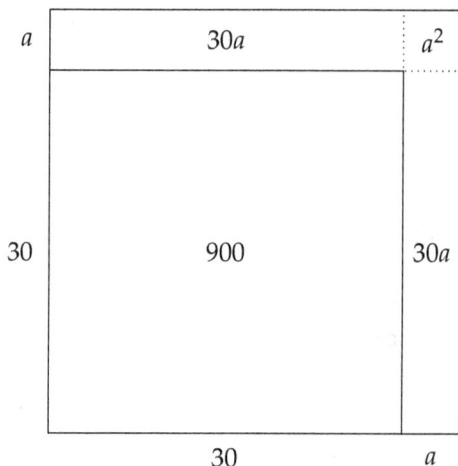

a	$30a$	a^2
30	900	$30a$
	30	a

Fig. 5.7   Liu's approach to $\sqrt{961}$

Liu knew that the area outside the 30×30 square was 61 and he knew (for the reason shown in Figure 5.7) that this was also equal to

$$30 \times a + a \times a + 30 \times a = 60a + a^2.$$

So

$$60a + a^2 = 61.$$

He didn't want to solve a quadratic equation – which presupposes the ability to find square roots – so he changed the problem to solving

$$60a < 61.$$

Now a can't be zero because that would mean there is no strip wrapped around two edges of the 900 unit square. So the only whole number value that a can have is one. How big is $60a + a^2$ with $a = 1$? Happily for me, it's 61. Which means that I have used up all of the area of the larger square. What's more, the side of this biggest square is

$$30 + a = 30 + 1 = 31.$$

So the square root of 961 is 31. Check it out by multiplying 31 by 31.

Now I need to extend the problem of finding square roots in two ways so that you can see how to cover all possible square roots. First, I'll do an example where the area is six digits long. And second, I'll do an example where the number I'm working with is not the square of a whole number. Having done that, by extension of these examples, you'll be able to find the square root of any number. Well, perhaps "any number" is a bit of an exaggeration. Once you can do it for a few, so that you know what Liu was on about, then I think you'll shoot back to a regular calculator/computer.

OK, then what is the square root of 184041? Let's go straight to the Liu diagram in Figure 5.8. The part of this number that I'll tackle first is 180000. There will be some biggest square that is less than 180000 and its numeral is of the form xy0000. That's got to be 160000, which gives us the side length 400 of the smallest square in Figure 5.8.

Now I have to find the area outside the smallest square. This is just $184041 - 160000 = 24041$. Using the strip idea from the 961 example, I'll look for a such that

$$400a + a^2 + 400a = 24041.$$

But I don't want to solve this equation, so I'll try to solve $800a < 24041$ instead. I'll look for a multiple of 10 that satisfies this inequality. The biggest multiple of 10 that does so is accidentally 30 (it has nothing to do with the 30 of Figure 5.7). But 30 is too big to solve the equation, because $400 \times 30 + 30^2 + 400 \times 30 > 24041$. That means that I have to go down to 20. So I can attach the strip shown in Figure 5.8 wrapped around the 400×400 square.

At this point I have a square of side 420, which has area 176400. I now have to cover $184041 - 176400 = 7641$. That leaves me needing a strip of

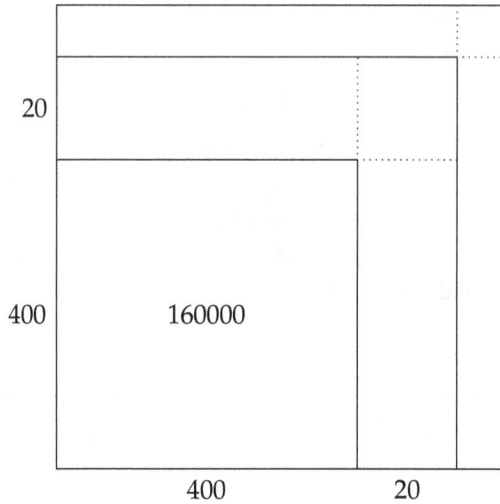

Fig. 5.8   The square root of 184041

width b where $2 \times 420 \times b + b^2 = 7641$. Using the inequality $840b < 7641$, gives $b = 9$ and this time $840 \times 9 + 9^2 = 7641$. I've got there with a square of side 429, telling me that $\sqrt{184041} = 429$.

There are a couple of new things here. One of these was needing three steps and two different strips. Another was having to backtrack on the 30 for the first strip. And note that I now seem to be taking steps in decreasing powers of 10. So we started with the first approximation using hundreds, the second using tens, and the third using units.

I now plan to look at a situation where the number involved is not a perfect square. The easiest way to do this is to add a small number to 184041, say 300. Then I can reuse a lot of arithmetic. Fine, so this time I'm battling with 184341. Everything will be exactly the same up to the production of the 429-sided square, so we'll start with that and continue to add strips. This new starting point is shown in Figure 5.9.

Now I have 300 to be covered by a strip of width a, so $858a + a^2 = 300$. In the last example I went down to units, so now I must be in the tenths. Calculation shows that 0.4 is too big, and 0.3 will do, for a. At this point my best answer with the last strip is going to be 429.3. But if I was anxious to get more decimal places, then I'd have to add another strip around the 429.3 square, and another, and ... I guess until I either got fed up with the whole thing, or I reached the accuracy I wanted.

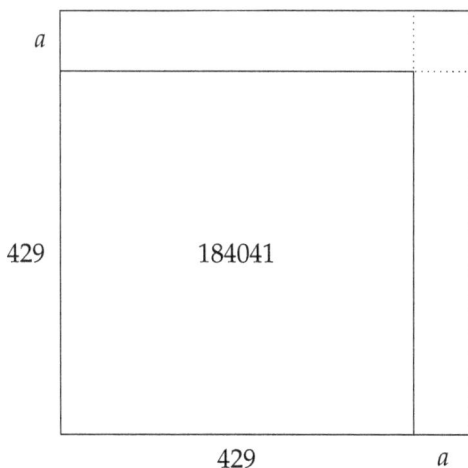

Fig. 5.9   The square root of 184341

You are now experts, so finally I want to say two things. First, this isn't Liu's algorithm – it was well known before he set about his Commentary. He provided the outline of a general proof, in much the same way as I have done here. However, working out the square root of a number that isn't a perfect square can be a pretty stoic effort. I like to think that he had a few companions who checked all the arithmetic for him.

Second, this shows that Liu knew the square of a sum. This is needed because the last few Figures were all about one square being the sum of a smaller square and a strip wrapped around two sides. Algebraically:

$$(a + b)^2 = a^2 + 2ab + b^2,$$

and verbally: the area of the biggest square = one of the smaller squares, plus the sum of the two rectangles, plus the other square. We saw this in Figure 5.3, already used to give a geometric proof of Pythagoras/Gougu.

5.5   Approximations to π

Finding π by length

Before I get to Liu, I want say a little about limits and infinite sums. In the discussion of Zeno's dichotomy paradox in Section 3.6 we saw that the infinite sum

$$\frac{1}{2} + \frac{1}{4} + \frac{1}{8} + \frac{1}{16} + \cdots$$

must equal 1, because each *finite* sum

$$\frac{1}{2} + \frac{1}{4} + \frac{1}{8} + \cdots + \frac{1}{2^n} = 1 - \frac{1}{2^n}$$

is certainly not *more* than 1, yet any number *less* than 1 can be exceeded by choosing n sufficiently large. In this case we say that 1 is the **limit** of the finite sum

$$\frac{1}{2} + \frac{1}{4} + \frac{1}{8} + \cdots + \frac{1}{2^n}$$

"as n tends to infinity", or that 1 is the value of the infinite sum

$$\frac{1}{2} + \frac{1}{4} + \frac{1}{8} + \frac{1}{16} + \cdots .$$

In this case we knew in advance what the value of the infinite sum would be, because we created it by cutting an interval of length 1 into infinitely many pieces.

The limit concept, and infinite sums, get more interesting when we use them to describe numbers not known in advance, such as π.

Now down to the real business. Archimedes and Liu never met, since they lived in different centuries. What's more, Archimedes' work on π almost certainly didn't reach China in Liu's time. Yet Liu discovered something very like the method that Archimedes used when he found his approximation for π (mentioned in Section 4.3). There were two small differences with Liu's method. The first was that Liu used only polygons inside the circle. Clearly the perimeter of an n-gon gets closer and closer to the circumference of the circle that circumscribes it as n gets bigger and bigger. This shows that the circle is the limit that the n-gons approach. So a polygon outside the circle isn't really needed, though it could give an upper bound to the estimate. The second difference I'll go into more detail in the next section. This, you'll see, is different from Archimedes' method because it uses the area of the polygons to approach the *area* of the circle.

However, Archimedes and Liu Hui used very similar polygons, and calculated their perimeters by repeated use of the Pythagoras/Gougu theorem. Each began with a hexagon and repeatedly doubled the number of sides by angle bisection. It is clear from the left part of Figure 5.10 that the perimeter of the inscribed hexagon equals six times the length of the radius, since the hexagon is made from six equilateral triangles whose sides equal the radius. So the first approximation to the circumference 2π of the circle of radius 1 is 6, giving the approximation $\pi \approx 3$.

After that, Liu Hui calculates the dimensions of each new polygon from those of the old with the help of Pythagoras/Gougu. The relevant

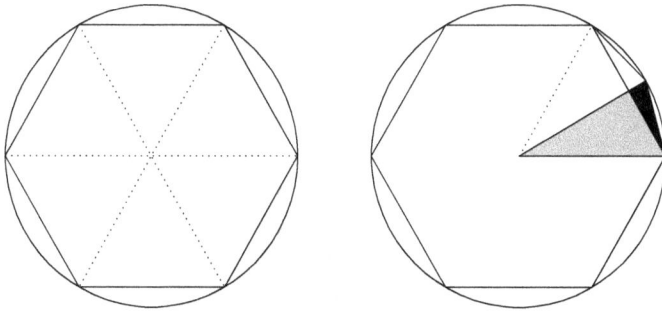

Fig. 5.10   From the hexagon in a circle, to the next polygon

right-angled triangles are shown in the right part of Figure 5.10. The radius (one) of the circle equals the hypotenuse of the grey triangle, and the shortest side of this triangle is $1/2$ (namely, half the side of the equilateral triangle), so we can find the remaining side by Pythagoras/Gougu. The latter side, plus the short side of the black triangle, equals the radius (one), so we can also find the short black side. We now have two sides of the black triangle, so we can find its hypotenuse, which is the side of the next polygon. A similar calculation takes us from *any* polygon in the sequence to the next.

Calculating lengths by Pythagoras/Gougu uses a lot of square roots, so you can see why Lui Hui was interested in square root algorithms.

It's clear from Liu's approach to π that he understood the concept of a limit, as did Archimedes. He also knew the idea of a recurrence relation, which is a way of finding each term in a sequence from previous terms. This is precisely what Lui Hui did in passing from each polygon to the next, calculating the new perimeter with the help of the old. We'll see more on recurrence relations in Chapter 9, when we look at Fibonacci's famous recurrence relation.

And Now for π by Area

Liu Hui had a crafty method for finding that the area of a circle was πr^2. I'll show it this way.

In Figure 5.11 I've cut a circle of radius r into a number of equal sectors. I have then arranged them side by side in a row. If the sectors are very thin their curved sides are almost straight, so the width of the row is almost the circumference of the circle, $2\pi r$. Also, they are almost triangles of height *r*. And as the sectors become thinner these estimates become more exact. Now if they *were* really triangles of height *r*, and the total width was $2\pi r$, their total area would be (since area of triangle $= 1/2$ base $\times$ height)

$$A = 2\pi r.r/2 = \pi r^2.$$

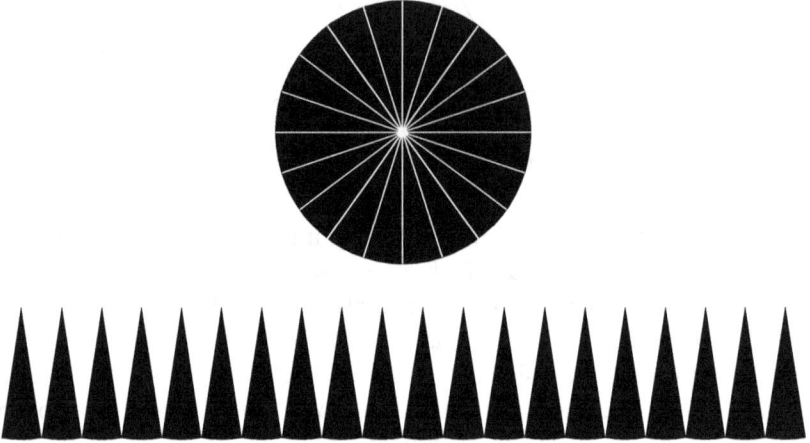

Fig. 5.11   Limits of sectors leads to πr^2

Now the total area of the sectors does not depend on how thin we slice them, so their total area (which is the area of the circle) cannot be other than πr^2 – since we can certainly get as close as we please to this number by slicing thinly enough. Then Liu Hui had a Eureka moment. Instead of approximating π using length, he approximated it by area. Knowing now that the area of the circle was πr^2, he built up the area of the circle using the pieces in Figure 5.10.

But he hadn't finished with limits. He now thought it might be useful to use the area of his polygons to find another route to a value of π. He might have gone straight to the area of an *n*-gon and worked in a similar way to the circumference approach in the last section. But what he did was to notice a few things. For example, he looked at A_n the area of a regular *n*-gon. And then thought about $A_{n/2}$, to see how much the polygons grew by. So he let D_n be this growth in area from $A_{n/2}$ to A_n, to give

$$D_n = A_n - A_{n/2}.$$

For example, A_6 is the sum of triangles like the one shown in grey in Figure 5.10, and D_{12} is the sum of triangles like the one shown in black.

Since the polygons approach the circle as closely as we please, A_n approaches the area of the circle of radius 1, which is π. Also, we have

$$A_{12} = A_6 + D_{12},$$
$$A_{24} = A_{12} + D_{24} = A_6 + D_{12} + D_{24},$$
$$A_{48} = A_{24} + D_{48} = A_6 + D_{12} + D_{24} + D_{48},$$
$$A_{96} = A_{48} + D_{96} = A_6 + D_{12} + D_{24} + D_{48} + D_{96};$$

So, instead of "growing" the subscript of A_n, producing areas that approach π, we can simply keep adding D_n terms. This led Liu Hui to describe π as the *infinite* sum

$$\pi = A_6 + D_{12} + D_{24} + D_{48} + D_{96} + D_{192} + \cdots.$$

Bearing in mind that each $D_n = A_n - A_{n/2}$, he could also write this infinite sum as

$$\pi = A_6 + (A_{12} - A_6) + (A_{24} - A_{12}) + (A_{48} - A_{24}) + (A_{96} - A_{48}) + \cdots.$$
$$(*)$$

Next there is a nifty trick that is used from time to time, where you seem to be putting a lot in, but most of it cancels out. Here the first A_6 is cancelled by $-A_6$ in the first bracket, the first A_{12} is cancelled by $-A_{12}$ in the second bracket, and we can keep doing this as long as we like. Liu Hui stopped after cancelling A_{96}, which left the series

$$\pi = A_{192} + D_{384} + D_{768} + D_{1536} + D_{3072} + \cdots.$$

This simply says that π equals the area of the 192-gon inside the unit circle plus infinitely many "correction" terms $D_{384}, D_{768}, D_{1536}, D_{3072}, \ldots$ which come from the thin black triangles added after constructing the 192-gon. So we can slowly improve the estimate A_{192} for π by adding more and more of the correction terms.

Now here is yet another clever part. Liu Hui found a way to speed up the correction process by noticing that each D_n is very close to $D_{n/2}/4$. So the infinite expression for π is approximately equal to

$$A_{192} + \left[(1/4) + (1/4)^2 + (1/4)^3 + (1/4)^4 + \cdots\right] D_{192}.$$

The summation inside the square brackets is one we've already seen in Euclid, and Liu Hui knew it too. It's a geometric series with sum $1/3$.

So now $\pi \approx A_{192} + (1/3)D_{192} \approx 3.1416$, which is an improvement on Archimedes. For more information, see Wikipedia, Liu Hui's π algorithm.

But I have to say that what Liu did in this mathematical argument is what mathematicians called "nice" or "pretty". He has put together some interesting observations on the relative sizes of the D_n relations, and come up with a good way to express π in an unusual way that suddenly turns into something quite simple. What's more this argument had never been used before, not even by Archimedes! It goes without saying that Liu didn't have the notation to put the argument the way I have. But that makes it even cleverer.

Another Use of the Subtraction Trick

I want to give another use of the difference method of equation (*) in the last subsection. I'm going to find the sum of the first 10 whole numbers. Bear with me as I set it up and as I go through it, because this nice mess is not just made for this single problem. I'll show you later the many things you can do with it. First look at some equations, which are instances of the identity $(n+1)^2 - n^2 = 2n + 1$ that follows from $(n+1)^2 = n^2 + 2n + 1^2$:

$$11^2 - 10^2 = 2 \times 10 + 1$$
$$10^2 - 9^2 = 2 \times 9 + 1$$
$$9^2 - 8^2 = 2 \times 8 + 1$$
$$\vdots$$
$$3^2 - 2^2 = 2 \times 2 + 1$$
$$2^2 - 1^2 = 2 \times 1 + 1$$

How do they give the sum of the numbers from 1 to 10? Well, they are set up for the subtraction trick. If I add up all the equations, the left side looks like

$$(11^2 - 10^2) + (10^2 - 9^2) + (9^2 - 8^2) + \cdots + (3^2 - 2^2) + (2^2 - 1^2).$$

Can you see that this has all been craftily composed so that almost everything cancels out, exactly as in Liu Hui's equation (*)? Except now we can get to the end, and the sum of the left-hand sides of the equations reduces to $11^2 - 1^2 = 120$. On the right-hand side, I've got twice something plus one. That comes because

$$(n+1)^2 - n^2 = (n^2 + 2n + 1) - n^2 = 2n + 1,$$

so the terms on the right-hand side sum to

$$(2 \times 10 + 1) + (2 \times 9 + 1) + (2 \times 8 + 1) + \cdots + (2 \times 2 + 1) + (2 \times 1 + 1)$$
$$= 2(10 + 9 + 8 + \cdots + 2 + 1) + (1 + 1 + 1 + \cdots + 1 + 1)$$
$$= 2(10 + 9 + 8 + \cdots + 2 + 1) + 10.$$

Putting all of this together

$$120 = 2(10 + 9 + 8 + \cdots + 2 + 1) + 10.$$

And this simplifies to

$$10 + 9 + 8 + \cdots + 2 + 1 = (120 - 10)/2 = 55.$$

Not a groundbreaking result, I'm afraid, but we can extend this to the sum of the first n numbers. Just start with the nth equation, and go down to the first again.

$$(n+1)^2 - n^2 = 2n + 1$$
$$n^2 - (n-1)^2 = 2(n-1) + 1$$

$$\vdots$$

$$3^2 - 2^2 = 2 \times 2 + 1$$
$$2^2 - 1^2 = 2 \times 1 + 1$$

Adding up the equations, I get

$$(n+1)^2 - 1 = 2[n + (n-1) + (n-2) + \cdots + 3 + 2 + 1] + n.$$

If I put $S_n = n + (n-1) + (n-2) + \cdots + 3 + 2 + 1$, then, since

$$(n+1)^2 - 1 = n(n+2),$$

I get

$$n(n+2) = 2S_n + n.$$

This gives me $S_n = n(n+1)/2$. To check, put $n = 10$ into S_n and see what I get: $S_{10} = 10(10+1)/2 = 5 \times 11 = 55$, as above.

General problem solved. Of course, Gauss did it a little easier (according to legend, see MacTutor, Gauss) but I'll leave that for another day.

But why have I taken so long on the lead up to all of this? Look at what we did. We used cancellation of *squares* to express the sum of the first n numbers in terms of something easy: the sum of n ones. In the same way, we can use cancellation of *cubes* to express the sum of the first n *squares* in terms of the sum of the first n numbers (which we now know) and the sum of n ones. And after that, we can use cancellation of *fourth powers* ... and so on. Here is how to cancel cubes in order to find the sum of the first n squares. Again we will start with 10 equations, but you will see how to extend the idea to n equations.

First you need to know that $(n+1)^3 = n^3 + 3n^2 + 3n + 1$. Then

$$11^3 - 10^3 = 3 \times 10^2 + 3 \times 10 + 1$$
$$10^3 - 9^3 = 3 \times 9^2 + 3 \times 9 + 1$$
$$9^3 - 8^3 = 3 \times 8^2 + 3 \times 8 + 1$$

$$\vdots$$

$$3^3 - 2^3 = 3 \times 2^2 + 3 \times 2 + 1$$
$$2^3 - 1^3 = 3 \times 1^2 + 3 \times 1 + 1$$

Now add everything up and you'll get $11^3 - 1^3$ on the left. But on the right, you'll have 3 times the sum that you are after, plus 3 times the sum we found earlier, plus 10. Tidy it all up and you have your answer for the sum of the squares of the first 10 numbers. Now you can find the sum of the squares of the first n numbers. Can you find the sum of their cubes?

5.6   Sea Island Mathematical Manual

When Liu was writing his version of the *Nine Chapters*, the nine problems that make up the *Sea Island* were contained in the *Nine Chapters*. But when Li Chunfeng assembled the Ten Mathematical Classics, he removed the *Sea Island* material to make a separate entity. This material, however, is included in the *Nine Chapters* translation of Shen *et al.* (1999).

The purpose of the new *Manual* is to help surveyors. Starting simply, we can measure building height if we can measure distance to the building, and use a gnomon with known lengths as in Figure 5.12.

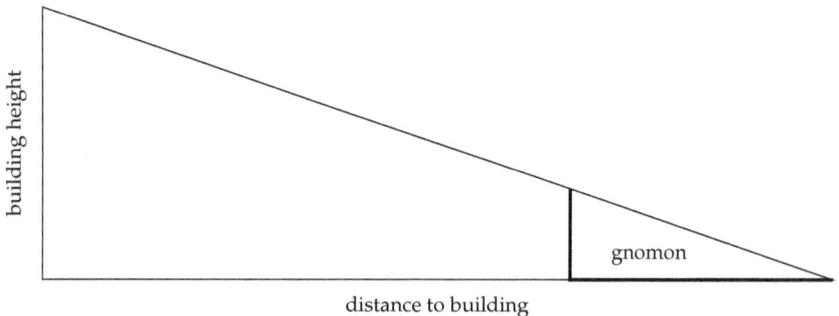

Fig. 5.12   Simple calculation of height

The calculation of the height of the building here is simply found by ratios since

building height/distance to building = ratio of gnomon lengths.

But what do you do if there is no simple way to find the distance to the building? Let's have a look at the first Sea Island Problem, shown in Figure 5.13 from the *Sea Island*.

Fig. 5.13 The sea island (from Wikipedia, Liu Hui)

Sea Island problem 1. Find the height of the highest point S on the island shown in Figure 5.14. (Here I'll assume that you can get a direct line of sight to the island's peak.)

Now I would probably have gone straight off looking for similar triangles, hoping it would all come out in the end. But Liu Hui had a theorem that gave a neater solution. Here is his theorem.

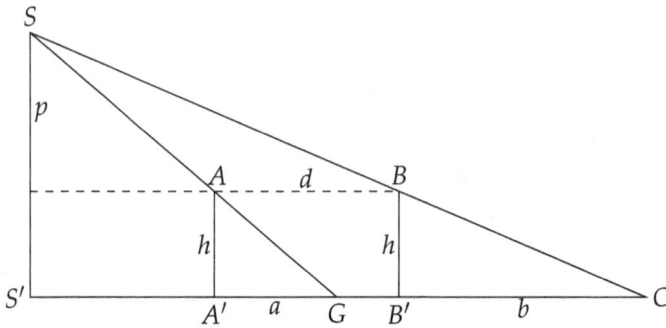

Fig. 5.14   Calculating the height without the length A'S'

Liu's Theorem. In Figure 5.15, the shaded rectangles have equal area.

Proof. Look at the six triangles in Figure 5.15. They make up three pairs of congruent triangles, each pair being halves of a rectangle.

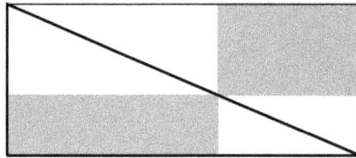

Fig. 5.15   Showing that the shaded areas are equal

(1) Upper and lower half of the big rectangle (each consisting of two white triangles and a grey rectangle).
(2) Upper and lower half of the larger white rectangle.
(3) Upper and lower half of the smaller white rectangle.

Each shaded rectangle equals a big triangle minus one from each pair of white triangles, so the shaded rectangles have equal area.   Q.E.D.

Now to find SS' we use the lower-case letters defined in Figure 5.16. From the diagram we can see that $(c + d)h = pb$ and $ch = ap$ by Liu's theorem (in two different rectangles, with diagonals SC and SG respectively). Combining those two equations we get $ap + dh = pb$ or $dh = pb - pa$. So $p = dh/(b - a)$. This gives $SS' = h + dh/(b - a)$. Rewriting the lower-case letters, we now see that
$$SS' = AA' + (A'B' \times AA')/(B'C - A'G),$$
and we've got the height we wanted.

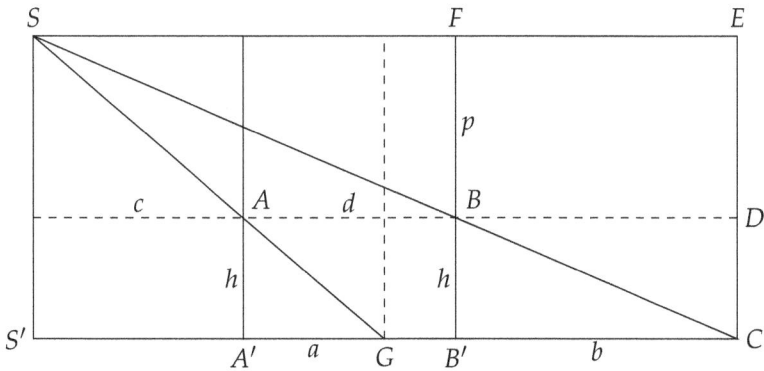

Fig. 5.16 Recalculating the height without the length $A'S'$

Note that the pairs of sides AA' and $A'G$, and BB' and $B'C$, would be gnomons of height h in a practical situation. This is essentially the case for most of the other problems in the *Sea Island*.

There is of course an assumption in the above situation, typical of basic surveying problems, and shown in Figure 5.13. Namely, that there is a flat piece of the mainland, near the island, where the points A', B', G, C could be found and from which a surveyor could see the peak.

Chapter 6

Chinese Mathematics After Liu Hui

After Liu Hui, Chinese mathematics continued to advance, often in directions where Liu Hui had pointed the way. I will look at three of these directions in this chapter:

- the **numerical computation** direction, which extended methods for calculating square roots to methods for calculating roots of arbitrary polynomial equations;
- the **combinatorial** direction, which introduced what was later called **Pascal's triangle** for calculating binomial coefficients;
- the **linear algebra** direction, adapting methods from simultaneous linear equations to simultaneous polynomial equations.

These advances, particularly the first and last, took Chinese mathematics to a level that was highest in the world at the time, at least in the theory of equations.

6.1 Numerical Solution of Equations

The square root process devised by Liu Hui (and a similar cube root process I didn't mention, where approximate squares are replaced by approximate cubes) is a way of solving polynomial equations *numerically*. The process described in Section 5.4 for finding $\sqrt{961}$, for instance, solves the equation $x^2 = 961$ by successive approximations. In this case we reach the exact answer $x = 31$ in two steps, but in general the process could continue indefinitely, as Liu Hui was well aware. In fact he said it could continue down to

those small numbers for which the units do not have a name.

Quoted in Martzloff (2006), p. 229

To avoid these "small numbers", square root algorithms tended to be used on large integers (such as 961) and to stop when they reached whole number units. The same preference for whole number computations is evident when we come to general methods for solving polynomial equations, which developed in the 13th century, notably in Qin Jiushao's *Shushu Jiuzhang* (Mathematical Treatise in Nine Sections) of 1247, and Zhu Shijie's *Siyuan Yujian* (Jade Mirror of Four Unknowns) of 1303.

Qin Jiushao uses a remarkably simple algorithm for solving an arbitrary polynomial equation by successive approximations. In modern terms, the method begins by finding (by trial and error) an interval between successive integers in which the equation has a root. Then it effectively "magnifies" the interval, by a change of variable, in order to confine the root to a smaller subinterval, and repeats until the root is approximated as closely as desired.

I'll illustrate the method on the equation

$$f(x) = x^3 + x - 1 = 0. \tag{1}$$

Substituting $x = 0$ and 1 gives $f(0) = -1 < 0$, $f(1) = 1 > 0$, so equation (1) has a root somewhere in the interval $(0, 1)$. Now we effectively expand this interval to $(0,10)$ by changing the variable to y, where $x = y/10$. In terms of the new variable the equation becomes

$$(y/10)^3 + (y/10) - 1 = 0,$$

that is,

$$g(y) = y^3 + 100y - 1000 = 0. \tag{2}$$

By trial and error we find $g(6) < 0$ and $g(7) > 0$, so equation (2) has a root in the interval $(6,7)$ of y-values, which is the interval $(0.6, 0.7)$ of x-values. If we are satisfied with one decimal place accuracy, we can stop there.

If not, we effectively move the interval $(6,7)$ of y-values to the interval $(0,1)$ of z-values by making the change of variable

$$z = y - 6, \quad \text{that is,} \quad y = z + 6.$$

With this change of variable the equation becomes

$$(z+6)^3 + 100(z+6) - 1000 = 0, \tag{3}$$

and it has a root in the interval $(0,1)$ of z-values.

We are now ready to repeat the process (by setting $z = w/10$) until the root is confined to an interval as small as desired. I just note a few things about the process:

- Computations are very simple, mainly evaluations of polynomials at integer values until we find consecutive integers at which the polynomial changes sign.
- The algebra is also quite simple, though it may require expansion of binomials to all powers present in the original equation. For example, in equation (3) we may want to expand $(z + 6)^3$.
- The process does not attempt to find *all* roots of the equation. Rather, if it finds *one* root, it can calculate that root to any required degree of accuracy.

This 13th-century discovery was one of the high points of Chinese mathematics. A similar process was not known in Europe until the early 19th century, when Ruffini and Horner independently discovered much the same thing. In some books it is known as "Horner's method".

6.2   Pascal's Triangle

As remarked above, the process for numerically solving equations involves expansion of binomials $(a + b)^n$ to all powers n occurring in the equation to be solved. To ease the calculation of these powers, Chinese mathematicians made use of the triangle of their coefficients that we know as **Pascal's triangle** of **binomial coefficients**. The rows of the triangle for $n = 0$ up to 8 are shown in Figure 6.1, followed by Zhu Shijie's picture of the same triangle in Figure 6.2.

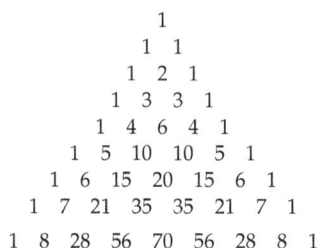

$$
\begin{array}{ccccccccccccccccc}
 & & & & & & & & 1 & & & & & & & & \\
 & & & & & & & 1 & & 1 & & & & & & & \\
 & & & & & & 1 & & 2 & & 1 & & & & & & \\
 & & & & & 1 & & 3 & & 3 & & 1 & & & & & \\
 & & & & 1 & & 4 & & 6 & & 4 & & 1 & & & & \\
 & & & 1 & & 5 & & 10 & & 10 & & 5 & & 1 & & & \\
 & & 1 & & 6 & & 15 & & 20 & & 15 & & 6 & & 1 & & \\
 & 1 & & 7 & & 21 & & 35 & & 35 & & 21 & & 7 & & 1 & \\
1 & & 8 & & 28 & & 56 & & 70 & & 56 & & 28 & & 8 & & 1 \\
\end{array}
$$

Fig. 6.1   Pascal's triangle up to row eight

Notice that the Chinese numerals in the triangle are similar to those discussed in Section 5.2, except that the zero symbol is used in the symbols for 10 and 20, and horizontal lines can denote multiples of 10. (Also, there seems to be an error in line 7, with a 34 that should be 35.)

圖　方　㮸　七　法　古

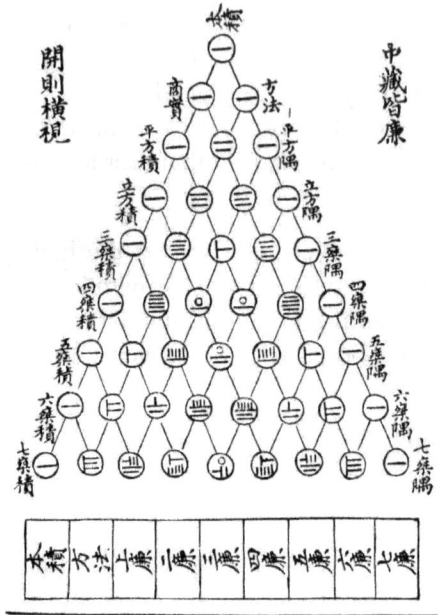

Fig. 6.2   The triangle in Zhu Shijie's *Siyuan Yujian* of 1303 (from Wikipedia, Pascal's triangle)

The Chinese mathematicians knew the simple rule for generating these coefficients – each is the sum of the two coefficients above it – but apparently not the formula $\frac{m!}{n!(m-n)!}$ for the coefficient of $a^n b^{m-n}$ in $(a+b)^m$. Pascal's contribution was to tie all these facts together, making it possible for them to be used in probability theory. (See Section 11.5.)

6.3   Linear Algebra

As we saw in Section 5.3, the method for solving linear equations in several unknowns that we know as "Gaussian elimination" was known in China about 2,000 years ago and later used by Liu Hui. The method "eliminates" all but one unknown, whose value is then immediate and can be used to find the value of each other unknown, one by one. A few centuries later, a similar method was used in China to solve a seemingly harder problem: eliminating variables from *polynomial* equations in several unknowns.

The *Jade Mirror of Four Unknowns* of Zhu Shijie, 1303, showcases the method by eliminating unknowns in up to four polynomial equations. In this way, Zhu Shijie arrives at an ordinary polynomial equation in one unknown, which can be solved by "Horner's method". Examples from the *Jade Mirror* may be seen (in English translation as well as Chinese) in J. Hoe's paper *The Jade Mirror of the Four Unknowns – Some Reflections*, which was published in Hoe (1978).

Eliminating three unknowns from four was perhaps as much as was feasible with the Chinese counting board techniques. But it was enough to show, I think, that it could be done with an arbitrary number of unknowns. In fact, the case of two polynomial equations in two unknowns is probably enough to hint at how elimination works with polynomial equations. Here is an example consisting of two quadratic equations in x and y:

$$x^2 + xy + y^2 = 1 \tag{1}$$

$$4x^2 + 3xy + 2y^2 = 3. \tag{2}$$

We pretend that the powers of y are "independent variables" and the powers of x are "constants", and proceed to eliminate the highest power, y^2, as in linear algebra. Namely, subtract 2 times equation (1) from equation (2), which gives

$$2x^2 + xy = 1, \quad \text{so} \quad y = \frac{1 - 2x^2}{x}.$$

Substituting this value of y in equation (1) gives a polynomial equation in x alone:

$$3x^4 - 4x^2 + 1 = 0.$$

Admittedly, if higher powers of y are present there will be more work. In that case, we need more than two equations in the powers of y – as many equations as there are powers of y, in fact. But such equations can be created, as the following example shows.

$$1 + xy + y^2 = 0, \tag{3}$$

$$3 + 2x^2 + y^3 = 0. \tag{4}$$

We first get a lower-degree equation in y; namely equation (4) minus y times equation (1):

$$3 + 2x^2 - y - xy^2 = 0. \tag{5}$$

We can now eliminate y^2 between (3) and (5), by adding equation (5) to x times equation (3). This leaves an equation linear in y, which we can solve for y in order to get an equation in x alone, as we did in the case of equations (1) and (2).

After Zhu Shijie

Hoe's paper on the *Jade Mirror* also has some interesting remarks on the course of Chinese mathematics after Zhu Shijie:

> Zhu Shijie is the last of the great Chinese mathematicians of the past whose works are extant. After him, mathematics in China went into decline, and in the 16th century, the Jesuits had to be called in to help in the reform of the calendar and other mathematical activities. The reasons for the rapid decline in mathematical activity have been sought in the economic and social conditions of mediaeval China, and in the prevailing philosophical attitudes of the period.
>
> However, the decline can also be partly ascribed to the kind of mathematics studied and the methods employed. We have already seen, for example, that the counting board notation led to a dead end, because it could not be extended beyond two dimensions. Moreover, the counting board procedures not only became increasingly complex, but, as in modern abacus computations, the intermediate steps are not retained, with the result that it is difficult to check one's calculations, short of re-working the whole problem. We saw also that the problems to which the techniques were applied were of little practical use. The result of all this is that once the method is lost, it was not only difficult to recover, but there was little incentive to do so.
>
> So that it is not surprising that, after the Mongol conquest, Chinese mathematics did not revive, although the tradition remained alive in Japan and Korea until the 17th century. It is possible too, that the semi-symbolic nature of the Chinese language itself acted as a hindrance to progress. For though it proved to be an advantage at first, giving the Chinese mathematicians a lead over their contemporaries, they were not led to try to develop a true symbolism as was done in the West, where the alphabetic, inflected languages really did make reasoning in the language of discourse difficult, and hence encouraged the search for a truly symbolic notation.

Of course, the decline of Chinese mathematics after the 13th century was not permanent. Today, Chinese mathematics is strong and is part of the worldwide mathematical culture. With modern communications, there is no longer a distinction between the mathematical techniques used in different nations.

PART 3

Fibonacci

Chapter 7

Life and Times of Fibonacci

Fig. 7.1   Fibonacci, or Leonardo of Pisa (Wikipedia, Fibonacci)

7.1   Background

Fibonacci and Pisa

If you know anything about Fibonacci (Figure 7.1) it is probably that there are things called **Fibonacci numbers**, which form the **Fibonacci sequence**

$$1, 1, 2, 3, 5, 8, 13, 21, 34, 55, 89, 144, 233, 377, 610, 987, \ldots$$

or that he made a terrible job of modelling rabbit populations. What you may not know is that he imported a new system of arithmetic into Europe, which means that we don't need to carry an abacus around with us all the time.

Pisa is different though. The most likely thing that you know about Pisa in Italy is that it has a leaning tower (Figure 7.2). But Leonardo of Pisa might be its second most well-known product after the tower. Leonardo's family was called Bonacci, so today he is known as Leonardo Fibonacci because Fibonacci means son of Bonacci. Actually, the name Fibonacci was not used until 1506 when a guy called Perizolo, who was a notary of the Holy Roman Empire, called him Leonardo Fibonacci. Anyway, Fibonacci was born in Pisa around 1170, just a few years before the tower was built.

Fig. 7.2   The leaning tower of Pisa (Wikimedia Commons, Leaning tower of Pisa)

Fibonacci travelled a lot in Algeria and it didn't take him long to realise the value of the Hindu-Arabic system of numerals that he found there. One of his major contributions to the world and to mathematics was to write a book in 1202 called *Liber abaci* (see Wikipedia, Liber Abaci). Incidentally, *Liber abaci* has nothing directly to do with the abacus. It simply means "book of arithmetic", but "arithmetic" in those days was done on the abacus, so "abacus" was understood as "arithmetic". The purpose of the *Liber abaci* was to show that arithmetic could easily be done *without* the abacus, by using written numerals.

The *Liber abaci* compared the Hindu-Arabic system of numerals to that of Roman numerals and made it clear that the Hindu-Arabic system was the superior. The book took the new place value system to Europe. From there, of course it has been adopted by most of the rest of the world. But the uptake took time. It was two centuries or more before the Hindu-Arabic system was used by most people, and the abacus remained competitive with written computation until both were overtaken by the electronic calculator in the 1970s.

After Fibonacci, other mathematicians have worked on the Fibonacci sequence. One was the 19th-century French mathematician Édouard Lucas (see Wikipedia, Lucas number). He varied Fibonacci's sequence by letting the first two numbers be 2, 1. This was part of his generalisation of Fibonacci's sequence to what are known as the Lucas sequences (Wikipedia, Lucas sequence). In addition, there are many new ideas that draw their inspiration from Fibonacci. In fact, there is a journal, *The Fibonacci Quarterly* (https://www.fq.math.ca/), that publishes articles on the Fibonacci numbers alone. Fibonacci is certainly a mathematical giant on whose shoulders many people have stood. In the 1980s there was an American rock band named The Fibonaccis (see Wikipedia, The Fibonaccis), but as far as I know they made no contribution to mathematics.

A Short Rave

Before continuing, I should say that I am becoming worried about the Eurocentricity of the maths that we teach in schools, especially English-speaking schools. In Part 2, I showed a Chinese perspective of maths and in this Part I'll look at work that was known in India well before it was known in Europe. And once again we will find that we should seriously think about naming concepts after those who really did something first. I hope, for example, that we can give more recognition to Indian mathematicians for their numerals and not just dismiss them with the words "Hindu-Arabic system". This rave of mine is supported by MacTutor (Indian Mathematics – Redressing the balance), where you can find

> Mathematics has long been considered an invention of European scholars, as a result of which the contributions of non-European countries have been severely neglected in histories of mathematics. Worse still, many key mathematical developments have been wrongly attributed to scholars of European origin. This has led to so-called Eurocentrism. The neglect of non-European mathematics is no more apparent than when studying the contributions of India.

(However, we may have to think of others too, such as the Mayans, to be completely impartial.)

Anyway, it turns out that one of the Indian mathematical discoveries is named after Fibonacci! So now I have to expose Fibonacci and the "Fibonacci" sequence, but before I do I should say that like all the other named mathematical things, the namee didn't call them by his name. In fact, the naming was usually invented after the namee had died.

So who knew about the sequence 1, 1, 2, 3, 5, 8, 13,...? Hints of it appear in India as far back as Pingala before 200 BCE, and a clear statement of the generating rule is in works of Gopala and Hemachandra, in the century before Fibonacci wrote *Liber abaci*. (See Wikipedia, Fibonacci sequence.) Very little is known of them, but it seems that Indian scholars were interested in finding all possible rhythms based on one-beat and two-beat notes. This explains why the numbers of these rhythms are precisely the "Fibonacci" numbers.

The rhythms would be easy to program but tedious to actually play, because most of them would lack any repetition or "groove". It would be like listening to Morse code; at least that is my opinion as a drummer.

Even supposing Indian drummers (or poets) are open to more varied rhythms, I think that most of the possible rhythms would sound awful. And therefore this Indian investigation was probably the result of mathematical curiosity – just as Fibonacci wasn't really interested in modelling rabbit population growth.

Anyway, Gopala and Hemachandra also knew the generalisation of the "Fibonacci" sequence: $a, b, a + b, a + 2b, 2a + 3b, 3a + 5b$ and so on. Obviously, the sequence I am talking about comes from letting $a = b = 1$. Other variations have been considered by Lucas and others. Very likely this sequence came to Fibonacci along with all of the Indian maths that found its way to North Africa. It was just part of the package. Then, without even thinking, Fibonacci put the sequence in his book as an example.

7.2   Fibonacci's Actual and Conjectured Life

Europe before Fibonacci's Birth

In Europe leading up to the time of Fibonacci, there was a peace that hadn't existed since the Romans had been in control of Europe. Consequently, there was a large increase in population that led to economic growth generally. As a result of a move away from the land there was an increase in towns and cities. This gave rise to more interest in education and many universities were established that have survived until today.

The improved economic situation also led to a development in the arts. See Wikipedia, High Middle Ages.

As for Italy, Sicily progressed to be an important kingdom under the rule of Frederick II. Elsewhere, city states grew. One of these was Pisa, but Venice, Genoa and Amalfi also became more powerful. However, these states didn't always interact peacefully. Trading was undertaken not only with other European centres but also with Northern Africa and to the east to India and even China. Routes such as the Silk Road, not only connected east and west by commerce, but religion and new ideas were also interchanged.

By the time of Fibonacci's birth, Pisa was a thriving city with contacts to many regions that were some distance away. People were spending less time in simply surviving, and a strong middle class, especially of merchants, was growing. Fibonacci's father was one of this class and he had been appointed to help the Pisan merchants in their trade with Algeria.

Before I start on Fibonacci's personal life, it is worth noting that a large number of images on the web illustrate questions that I raise. You might find it useful to look at these. They will in many ways be more useful than what is written below. However, beware, if you search for 1200 AD, say, you may find yourself looking at an advertisement.

Fibonacci's Reputation

Most of the details that I have written in this section are based on reading the Background in Hughes (2008). This is a very interesting book and well worth reading if you are interested in Fibonacci and his works.

Hughes (2008), p. vii, starts by saying

> Leonardo of Pisa (Fibonacci) is the preeminent European mathematician of the Middle Ages. Besides being, in his own right, a creative mathematician, he was also a compiler of and an evangelist for the mathematics he believed was needed for his times.

This is not just a modern view. Frederick II thought so too, because he included Fibonacci in the group of thinkers he relied upon for advice.

Fibonacci's Personal Life

I must apologise for the Fibonacci portrait used at the start of the chapter. It is almost certainly a poor likeness, made centuries after his death, and well before smart phones had come on the scene. For the same reason, we know very little about Fibonacci as a person. In different places we find some statements about what he was like. But was he tall or short? Fat or thin? Would he have been a good companion if you were together with him on a desert island? The problem is that there are very few records about him in anything that can be dated from his lifetime or shortly thereafter. The only things that are said about his personality were written more than a hundred years after Fibonacci's death and so are unreliable.

Two things might be worth quoting for the insight they give. The first is from *Liber abaci* so that we can rely on that, and the second is from an official decree from the Republic of Pisa. Both quotes can be found in the MacTutor biography of Fibonacci.

> When my father, who had been appointed by his country as public notary in the customs at Bugia acting for the Pisan merchants going there, was in charge, he summoned me to him while I was still a child, and having an eye to usefulness and future convenience, desired me to stay there and receive instruction in the school of accounting. There, when I had been introduced to the art of the Indians' nine symbols through remarkable teaching, knowledge of the art very soon pleased me above all else and I came to understand it, for whatever was studied by the art in Egypt, Syria, Greece, Sicily and Provence, in all its various forms . . .
>
> the serious and learned Master Leonardo Bigollo . . .

Bigollo was a name that Fibonacci used from time to time. In Tuscan this meant "traveller". On the other hand, it can also mean "loafer", so some may have thought he spent a lot of time on things that were pretty useless. Further, I should say that Bugia was the capital of an autonomous province of the Almohad Empire; see Wikipedia, Almohad Caliphate. Bugia is now called Béjaïa, and is a little to the east of Algiers.

So we know that he was born in Pisa and moved to Bugia, North Africa, as a boy, where he studied accounting. In doing so, he learned about the Hindu-Arabic numerals and their arithmetic. It is also known that despite learning arithmetic he was soon very interested in mathematics. "I was so delighted with this knowledge that I preferred it to all other subjects." Accounting was chosen for him so that he could assist his father in his father's work as a notary. Incidentally, this was not a one-off appointment. Before 1100 CE, there was a commercial treaty between Pisa and the Maghrib, a Muslim region from Alexandria to the Atlantic Ocean. This important treaty was renewed regularly, even when in times of trouble. Under the treaty, the notary was to formally keep records of imported goods that were under his care and ensure that the appropriate taxes were paid. Presumably he encouraged two-way business as well. As this meant that the notary needed to know Arab law, he was almost certainly fluent in Arabic – as was Fibonacci himself, since he was schooled in Bugia. (See Hughes, 2008, p. xxi.) Also, in writing his book *De practica geometrie*, he would have read Euclid's *Elements* – in Arabic.

Fibonacci also travelled widely throughout the Maghrib as well as along the Mediterranean Sea as far east as Constantinople. We know that his father had a very responsible job in Bugia and so he was probably well paid. So Fibonacci was brought up in comfortable surroundings.

Fibonacci also did valued work. In Pisa he taught the local merchants the Hindu-Arabic number system and this was recognised by King Frederick II of Sicily with a permanent salary (see Section 7.4 below and Wikipedia, Frederick II, Holy Roman Emperor). This may well have made him financially independent for the rest of his life or at least supported his work in Bugia. My point here is that Fibonacci was probably a wealthy man.

First, from the fact that he wanted to teach merchants something that would advantage them, especially over the local rival city states in Italy, he taught them face to face. What's more he wrote books to teach various things. So I suspect that Fibonacci liked to teach and he was probably good at it or Frederick II wouldn't have handed over the salary. And merchants who had been calculating for years with the abacus probably needed an inspiring teacher to get them to change. His teaching approach can also be seen in his book where examples play a key role.

To get personal and public ideas I have looked at what is known of wealthy people generally around 1200 CE, both to the north and south of the Mediterranean. Hopefully what I produce below will give a picture of how Fibonacci might have lived. The extra thing that you need to bear in mind about Fibonacci is that he lived in two distinct cultures on two different sides of the Mediterranean.

Where Might Fibonacci Have Lived?

He would have lived in a stone building in both countries, as opposed to the wooden buildings or tents of the poor. A European house had a number of rooms including a kitchen, with a spit and hooks to hang pots on. Richer men would have a minstrel's gallery. There were also bedrooms for the women of the family. Almost certainly he would have had servants to keep the building clean, prepare food, wash clothes and so on.

In Islamic regions, there was much emphasis on decoration. Houses might include tiled areas as well as richly embroidered carpets.

It is worth noting that little furniture has survived from before the beginning of the 14th century. There are two likely reasons for this: first that the furniture was made of wood and second that little was actually made. It seems that even rich people, when moving to another residence, packed up all of their furniture onto a cart and moved it to the new venue.

During the Crusades (1059 to 1291, see Wikipedia, Crusades), soldiers had seen finely decorated residences around them where they fought and so there was a consequent improvement in their own castles or manor houses on their return. Adding to this push for more luxurious living, the women left at home in this period had been in charge of the management of the buildings they lived in and had taken the opportunity to liven up their previously drab surroundings. Unfortunately, no one seems to know whether Fibonacci had a wife or children, and if so, whether they went back and forth between Pisa and Bugia with him.

What Did He Eat and Drink?

In Pisa, Fibonacci would have access to wine, beer and water. But the peasants had less access to alcohol and their water might have been of poor quality. Wealthy men had the best wine that may have travelled from some distance. They would have made a show of it to guests. Milk was also available but this was predominantly drunk by children and the elderly.

Of course, drinking alcohol was forbidden by Islam. This might have led Fibonacci to an ethical problem. Should he have wine when his local friends weren't able to? As far as food was concerned, Islamic countries had a range of halal food where meat was killed in the proper fashion.

Meat was rare among the European commoners, but everyone had bread. Fish was accessible to all only if they lived near water. Of course, wealth enabled preserved fish to be brought some distance. Surprisingly, the richer citizens didn't eat green plants. This was because they thought it brought illness. The poor ate it regardless.

What Did He Wear?

In Europe, fashion changed little from the end of the Roman Empire up to the early 14th century. Perhaps women tackled the boredom of this by the way they decorated their clothes. Both sexes though had pretty much the same clothing with the only variation being in length.

Starting from the inside, people wore shorts called braies as undergarments. These were tied at the waist. Men wore shirts that hung down to the hip while women wore them longer (as they did in ancient Greece). These shirts seemed to be much like nighties that had a hole for the head with a slit to make it easier for them to be put on. A similar piece of clothing with longer sleeves, the tunic, was worn on the outside. Again, men had the short version that hung to the ankles while the women wore the tunics to the ground. More than one tunic might have been worn at a given time, presumably depending on the weather.

Loose "stockings" completed the ensemble for both sexes. Shoes were much the same too. They were made from leather stretched over a wooden mould. They tended to be pointed and turned up. This fashion lasted until the 16th century. Because the mould was crucial, shoemaking was a carpenter's job. Short boots were made for working.

The women did break the monotony with their hair. This was worn long and parted down the middle. From there it was plaited. Generally a small hat was attached. Men meanwhile, could do what they liked with their hair, being short or long or shaven didn't seem to be a problem. Beards were also allowed.

At least some of the Islamic clothing styles were similar to European. Both sexes wore simple tunics and a loose covering of the head. The women again wore longer garments than the men. But there are also things that were due to Islam itself, in particular, the veil worn by women. It's worth noting that veils were often worn by women in some areas before Islam arrived. The four types, hijab, niqab, chador and burqa, are described in Wikipedia, Hijab.

How Did He Travel?

Well, by land and water but not by air around Fibonacci's time. In Europe, on land there were feet, horses and carts. Not surprisingly, horses were the fastest, but carts carried the most goods. The basic problem was the roads. After Rome lost its power, sealed roads were rare except in and close to cities. The result was that rain and erosion were constant hazards to carts and inevitably their progress was often retarded.

Canals were built but were not yet at their prime. Sail was the power at sea and provided reasonable transport in the Mediterranean.

Camels were added to the list of transport possibilities in Africa and on to India and China. As commerce grew and merchants travelled to buy and sell, caravans became common. These could contain a large number of people. This attracted doctors, entertainers, and so on. Fibonacci travelled quite a bit in his work and may well have travelled in caravans for parts of his journey. Because of the variety of people in a caravan he wouldn't have been out of place in Italian or Algerian garb.

But did Fibonacci behave as a Pisan when he was in Italy, and an Algerian when he was in Algeria? Most likely, wherever he was he still had a foot in the other camp. Both authors of this book have lived for substantial periods in two different countries and they carry some ideas and speech and preferences of their "adopted" country. These authors would have tried to fit into the "new" country by learning the local language to some extent, driving on the other side of the road if appropriate, and so on. Fibonacci went further than just a few words of Arabic. Did he ride a horse in Pisa and a camel in Africa? Did he wear Islamic clothes every day there and use European dress when he was presenting to officials or on other formal occasions? I have no way of knowing.

7.3   Other Fibonacci Books

As far as is known, Fibonacci produced six books. These are

- 1202, *Liber abaci* (second edition in 1228);
- 1220, *De practica geometriae*;
- 1225, *Flos*;
- 1225, *Liber quadratorum*;
- ????, *Di minor guisa*;
- ????, Commentary on Book X of Euclid's Elements

The last two books mentioned here are now lost. Their existence is known because they have been referred to by other authors. In this section we will make comments on the other four books.

Liber Abaci

Fibonacci introduced the Fibonacci sequence to reinforce the practice of addition in Hindu-Arabic numerals. But there is more to the *Liber abaci* than Fibonacci numbers. First there is a big collection of problems aimed at merchants. Presumably this was because Fibonacci thought they would be the largest group to benefit from the Hindu-Arabic system. Unfortunately for Fibonacci, merchants were very slow to change and many of them still used the abacus centuries later.

The *Liber abaci* also contains other mathematical ideas. For instance, simultaneous linear equations, perfect numbers, problems involving the Chinese remainder theorem (Wikipedia, Chinese remainder theorem), $\sqrt{10}$ and approximations to it, and summing arithmetic and geometric series. There are also many problems included that originated in China. Here are three examples.

(1) A spider (in some versions, a frog) climbs a certain distance up a wall each day and slips back a fixed distance each night. How many days does it take him to climb the wall;

(2) A hound whose speed increases arithmetically chases a hare whose speed also increases arithmetically. How far do they travel before the hound catches the hare;

(3) Calculate the amount of money two people have after a certain amount changes hands and the proportional increase and decrease are given.

These problems attempt to get general solutions. For example, in the spider problem, you have to say how far the spider goes up in the day and down in the night. In addition to this you need to say how far from the top of the well the spider is to start with.

De Practica Geometriae

You probably guessed that this is a book on geometry. A thorough discussion of it can be found in

```
https://old.maa.org/press/periodicals/convergence/
mathematical-treasure-fibonacci-s-practica-geometriae
```

Continuing Fibonacci's penchant for problems and mathematics, this book contains problems and theorems based on Euclid's work (Chapter 2), including his work on division of figures into equal parts. There is also practical advice for surveyors such as how to find the height of a building using right-angled triangles, not unlike Liu Hui's *Sea Island*. (If you are keen on geometry, see the English version of Hughes (2008).)

Flos

The word "Flos" means "flower". I'm not sure why this title was used. And here we get back to good King Frederick II. The king invited Fibonacci to tea one day in 1220 and had Johannes of Palermo challenge Fibonacci to solve problems that Johannes had collected. Apparently, the problems were not original to him but were from Chinese sources.

One of these problems was to find a root of the equation

$$x^3 + 2x^2 + 10x = 20.$$

Fibonacci solved this using an approximation method. Strangely, he expressed his solution in the "sexagesimal", base 60, system introduced by the Babylonians. You would expect that Fibonacci, who was trying to convince Europe of the value of Hindu-Arabic numbers, would have used that system to express the answer. However, we can't talk – we also use base 60 when using minutes and seconds to measure both angles and time.

Also worth mentioning is that in Fibonacci's time it was not known that cubic equations could have just one real solution (which is the case for the equation above) or three real solutions.

Liber Quadratorum

This is considered to be Fibonacci's best mathematical book and contains new results. Unfortunately it has not been given the attention it deserves, though a translation into English of the *Liber quadratorum*, by Sigler, is in Fibonacci (1987).

It delves more deeply into mathematics than *Liber abaci* does. The title means the book of squares and it is essentially about number theory. One of Fibonacci's results shows that the squares can be found iteratively by using a recurrence relation that adds consecutive odd numbers to previous squares. For example, 1 is the first square. Now adding 3 to 1 gives the next square 4. Then adding 5 to 4 gives 9, the next square. So, using 1, 3, 5, 7, 9, 11, ... it follows by induction that all squares can be constructed. If you know a little algebra about general squares, you may be able to see how Fibonacci discovered this.

Fibonacci solved the problem of finding squares $x^2 < y^2 < z^2$ such that $y^2 - x^2 = z^2 - y^2$, and conjectured that in this case $y^2 - x^2$ is not a square, a result that was later proved by Fermat.

One other result involves Pythagorean triples (see Chapter 1): the whole number triples (a, b, c) such that $a^2 + b^2 = c^2$. This implies that

c is the length of the hypotenuse of a right-angled triangle whose other sides are a and b. To find these triples Fibonacci takes some odd square, say $3^2 = 9$, then adds 1, 3, 5, 7, the odd numbers less than 9. Since $1 + 3 + 5 + 7 = 16$, a square, we have $a^2 = 9$ and $b^2 = 16$. But also, $c^2 = 16 + 9 = 25$, giving $c = 5$. As a result, we have found the Pythagorean triple $(3, 4, 5)$. The method works for any odd square, because the sum of all odd numbers up to any point is always a square.

7.4 Historical Events in Fibonacci's Time

During Fibonacci's life, Italy was a collection of city states. Pisa was one of the maritime republics and there was antagonism between it and Genoa. Pisa was at the height of its power in Fibonacci's lifetime. It controlled the Western Mediterranean and claimed Corsica and Sardinia.

Frederick II was the Holy Roman Emperor at this time, so first we must consider the Holy Roman Empire. What was this institution? It covered western, central and southern Europe and existed from the early Middle Ages until the Napoleonic Wars in 1806, reaching its peak in the middle of the 13th century. After this period, its strength and extent diminished.

The Empire was founded on the idea of the ancient Roman Empire and the Emperor held equal power with all European monarchs. After his 1806 victory, Napoleon annexed what was left of the Holy Roman Empire which soon became known as the German Confederation. This eventually produced modern Germany.

Frederick II was quite a guy who was known by the Latin phrase "stupor mundi", which means "astonishment of the world". His overachievements are listed below.

1198: He began his rule of Sicily after the death of his father Henry VI. Frederick was about three at the time and was expected to reign the kingdom with the help of his mother, Constance. She died the same year and Pope Innocent III took the role of his guardian. But from 1198 until 1208 Sicily was run by others, including Frederick's tutor Cencio, who had Frederick imprisoned in the palace at Palermo. There Frederick learnt five languages: Latin, Greek, Arabic, Provencal, and, of course, Sicilian. Later he added Middle High German to this list.

Cencio's temporary seizure of power was aided by Genoese ships. This may be why Frederick later supported Pisa in its rivalry with Genoa. In 1208 he had to gain real power over Sicily. This was aided by the Pope who arranged a marriage for him with the widow of the King of Hungary.

She was twice the age of Frederick but she had good contacts who would support Frederick.

1212: He became King of Germany with papal support. But he wasn't crowned as Holy Roman Emperor in Rome until 1220 because of the many attempts to take over various regions of the Empire.

1220: He was made King of Italy and Holy Roman Emperor. However, he spent most of his time until 1236 either in Sicily or prevaricating about going on a crusade or actually going on one. He had said that he would go on a crusade when he was elected king of the Romans. He sent forces, yes, but didn't quite make it himself. Basically, there were troubles at home and he needed to get them sorted.

1225: For missing the Fifth Crusade Pope Gregory IX excommunicated him. Then there was the Sixth Crusade. Frederick II said he would sail to Jerusalem, but an "illness" occurred that prevented his departure. Finally, Frederick II set sail and led forces in the Crusade. If you think about it, an excommunicated Catholic, no matter who that was, couldn't actually lead a crusade. Frederick was excommunicated again.

The crusaders who were in Jerusalem didn't particularly want to proceed with Frederick at their head. So Frederick devised another scheme. He made a deal with the Muslim leader al-Kamil. This was good for the Muslim leader because he was under threat by other Arabian forces and al-Kamil retained parts of Jerusalem that were important for Muslims. It was good for Frederick because he gained the kingdom of Jerusalem, even if it was now a relatively smaller strip than existed earlier (Figure 7.3) and because he had stopped the fighting. It was not so good for many of the other leaders of the crusade. They could see that Frederick had gained ground, but didn't see the recovery of the Holy Land from the Muslims, which was, after all, what they were there for. In 1244, Jerusalem was retaken by the Muslims! In 1225 Fibonacci was asked to visit Frederick when his court went to Pisa.

1229: From here on, Frederick seemed to have his work cut out with incursions all over the Holy Roman Empire. One of his problems was from non-Europeans. The Mongol Empire attacked and defeated the forces of Hungary and Poland. Frederick told his allies in Europe not to engage with the Mongols and urged them to store food and arms. The Mongol leader Batu Khan demanded that Frederick submit, but he ignored the demand. Eventually, for reasons that are not completely clear, the Mongols were satisfied with their looting and they withdrew. See Wikipedia, Mongol invasion of Europe.

Fig. 7.3   Jerusalem during later crusades (see Wikipedia, Baron's Crusade)

1250: Frederick died in December that year. In his last years he always seemed to have opposition in some way that was both political and military. Generally speaking, he always won, either by clever plotting or by strength of arms. He might have been unhappy that his line did not survive. The Holy Roman Empire declined after Frederick's death in 1250 and his kingdom quickly broke up. This meant that he wasn't available to help the Empire's former ally when things hotted up between Pisa and Genoa in the second half of the 13th century. In 1284, the conflict led to the battle of Meloria (Wikipedia, Battle of Meloria) in which Pisa was defeated. After that its influence faded.

The Crusades lasted from 1095 to 1291 but essentially just involved Jerusalem. They were religious wars resulting from the invasion of

Jerusalem by Muslims. Essentially, the Popes at various times felt that Christian forces from Europe needed to win back the area because of its importance to Christianity. Altogether there were eight major Crusades and they had a variety of success.

Genghis Khan

He lived between about 1160 and 1227. Starting by uniting the Mongolian confederations, Genghis Khan and his Mongolian army conquered China and a large part of central Asia. In comparison with the Crusades and the Pisa/Genoan conflict, this was a real war. As a result, the Mongol Empire that was finally established lasted from 1206 to 1638. At its peak it was the biggest empire in world history that could be walked from one end to the other without getting your feet wet (Figure 7.4).

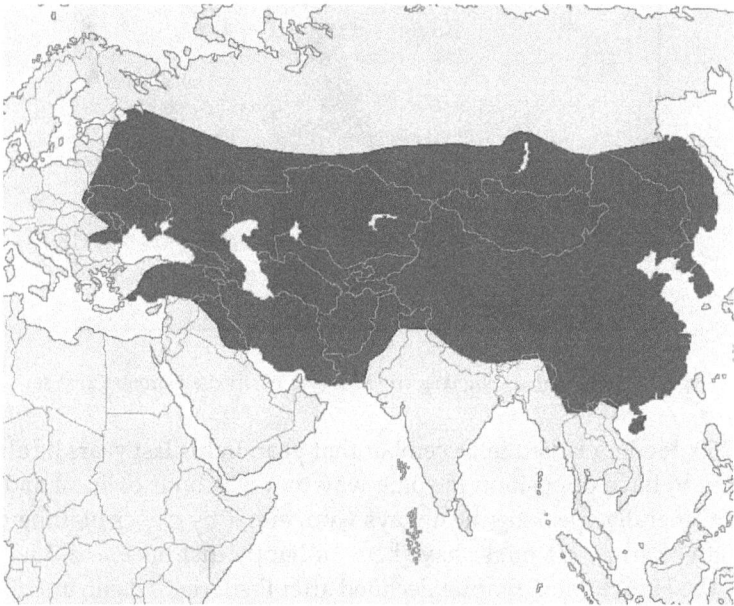

Fig. 7.4   Extent of Mongol empire (from Wikipedia, Mongol Empire)

The empire covered 23 million square kilometres, which is larger than the present Russia, and significantly larger than the empire of Alexander the Great (Chapter 1). (See the animated map in Wikipedia, Mongol Empire.)

Chapter 8

Mathematics before Fibonacci

8.1 Roman Numerals

Basic Numbers

Around about 400 CE the Roman Empire fell. Fibonacci was born in 1170. I note these two dates because although it may not have been surprising that the Italian region still retained and used Roman numerals when Fibonacci was alive, it is certainly surprising that the whole of Western Europe still did. Anyway, as a result, Fibonacci learned to write numbers using Roman numerals. But actual calculation, of sums and products and so on, didn't use Roman numerals. What Fibonacci and all the merchants and record keepers of his day did was to use a calculator! They worked out the results of the basic operations using a counting board or an abacus. Then they wrote down the answers on the document, where the answer was needed, in Roman numerals.

In the next few sections, I'll give you some idea of how the Roman system worked, covering the numerals and their meanings. Counting to ten in Roman numbers, Fibonacci would have been used to writing out I, II, III, IV, V, VI, VII, VIII, IX and X.

When Fibonacci used Roman numerals, he was standing on the shoulders of some unknown giants who developed Roman numerals. It was not unusual to use letters of the alphabet as numbers in those days. The ancient Greeks did this too, but less efficiently. The Roman way of doing things is more efficient and better than just using single strokes. The X was used because the Romans used a decimal system, possibly because we have 10 fingers (including thumbs) on our two hands. However, they didn't go as far as having a place-value system like the one we currently use.

The Romans avoided a large number of I's by inserting V for five and putting I before V or X to save all the ones they would have had to write if they had written IIII or VIIII. Despite this, IIII and VIIII were commonly written in ancient Rome and in later years if someone found them more useful than IV and IX. The I in IV and IX is subtracted from the V and the X, which is a little awkward, but I guess if you grew up with it you got used to it.

Having seen the value of V and the neat(?) advantage of the subtraction of I, the Romans consistently continued with the same basic idea for the tens, where they used X, XX, XXX, XL, L, LX, LXX, LXXX, XC. The new letters of L and C stand for 50 and 100. You can see how the implicit decimal system continued. Using the "when you are on a good thing stick to it" principle they moved on to the hundreds having only to introduce D for 500 and M for 1000. Then for 400, 600 and 900, they wrote CD, DC, and CM, respectively.

At this point you can make any number from 1 to 1000 in the way you usually do it with decimal digits, but the "digits" here are taken from the sets we've talked about above and show below.

Hundreds	Tens	Units
C, CC, CCC, CD, D,	X, XX, XXX, XL, L,	I, II, III, IV, V
DC, DCC, DCCC, CM	LX, LXX, LXXX, XC	VI, VII, VIII, IX

Hence 368 is just CCC LX VIII. Here I have left space to emphasise the position of the letters that represent hundreds, tens, and units. I won't do this from here on. To give you a few more examples, DCXXXIX = 639 and CMXCIX = 999.

Now you may think you know of a better way to write 999 in "Roman". Write it down. It's so much more efficient than the way I have written it. But IM is wrong. Or at least, it's not the way the Romans of 200 BCE would have done it. They kept fast to putting together the exact numbers of hundreds, then tens and then units.

Now I should write out the thousands for you. Well, I have and I haven't. I hope you are expecting a new letter for 5000, but I'm afraid that we are stuck with and going to use V. Obviously V still means five, so V would be ambiguous if it was left as it is. So, for 5000 the savvy Romans put a bar over the top of the V to get $\overline{V}$. That means we can do all the thousands from 1000 = M up to 9000 = ... Oh but what is 10,000? I'm sure that you can work that out. To see what the Romans used to use for higher numbers there are lots of web sites that will help you.

Written numbers were often just put on the page with the sum at the end. This is exactly the same as the receipts you get from your local supermarket. For a long time though, Roman receipts were produced via an abacus, but they never found an automatic way to do arithmetic completely on paper or papyrus.

Lingering Roman Numerals

Before we talk about *numerals*, which are symbols for numbers, we really should talk about number *words*, which are used to describe numbers at the most basic level. In any maths lecture, no matter how abstract, you are likely to hear words such as "fifty-seven", because you can hardly say 57 otherwise. Even in print, number words are often preferred to number symbols. Most publishers prefer "the nineteenth century" to "the 19th century", and the *New Yorker* magazine goes to extraordinary lengths to write numbers in words. For example, in the article "The Mountains of Pi" in the issue of 2 March, 1992, we are told that the number of decimal digits of π then known was

two billion two hundred and sixty million three hundred and twenty-one thousand three hundred and thirty-six.

(Despite that, I highly recommend the article.)

If number words are so persistent, it is no wonder that the much more compact Roman numerals are persistent too. In Chapter 7 of the *Liber abaci* we can actually see Fibonacci vacillating between number words and Roman numerals as he tries to talk about fractions. He mentions "twelfths" but cannot bring himself to say "eighteenths" – instead he says "XVIIIths".

At the start of this section, I expressed surprise that Roman numerals had lasted up to Fibonacci's time. But of course, we still use them. For instance, at the beginning or end of a movie you may see MMXXII. You can now tell your mates that this movie was released in 2022. And you can find out how old the old programs on TV are. That's one of the uses we have for Roman numerals. Here are some more.

(1) Kings, Queens and Popes, for example, use Roman numerals to distinguish between past people with the same name. Some examples are: King George VI, Queen Elizabeth II and Pope John XXIII;
(2) Euclid's book *Elements* has Books I, II, ..., XIII;
(3) Important games such as Olympics (XXI Olympics) and Super Bowls (Super Bowl LVI);

(4) Parts of the periodic table; and

(5) Some clocks have the Roman numerals I to XII instead of the Arabic numerals that we usually use.

8.2 Zero History

Zero

Before we start, *A history of Zero* at

`https://mathshistory.st-andrews.ac.uk/HistTopics/Zero/`

is well worth reading as it gives some more information and ideas on the subject. For example, zero didn't grow and develop gradually over the ages like many things in mathematics do. Zero appeared in some format then slipped away and was forgotten until it came back again somewhere else. This is at least partially due to the way that arithmetic was used. In the early stages people were interested in finding numbers of actual things, goats and armies and so on. We saw this with Diophantus. He did not consider negatives as an answer for a number of animals, because the concept of negative animals made no sense. And if you were trading, you didn't want or need to consider zero goats. (Except, of course, with pleasure after you had sold all the goats you had for sale.)

Being a sophisticated reader, you will know all about zero. It will come as no surprise to you that it can be used as a place-holder in *numerals* so you won't have to worry about whether 318 really means 318 or 30000108. And you will also be clear that zero is the *number* 0, which seems pretty useless until I tell you about it a little further on (see Section 9.6). Why would you want to know that 454545 + 0 is still 454545? But you, the reader, are probably hesitant to divide 7898 by zero, because, hopefully, your primary teacher told you that was unwise to try.

I suspect that you haven't given too much thought to Mayans, see the next section, who had a great civilisation from about 1500 BCE and lived in the area of central America containing Belize, El Salvador, Guatemala, Honduras and parts of southern Mexico including the Yucatán Peninsula. They flourished until about 900 CE, and their culture finally disappeared when they were conquered by the Spanish in the 16th century. However, in their prime the Mayans knew all there was to know about the place-holder value of zero, written as a shell shape. They certainly distinguished between 318 and 30000108.

It took other parts of the world a long time to catch up. The problem was that the Mayans couldn't get an email to their mates in China, Europe, India and elsewhere and explain it all to the rest of the world. We don't know how the Mayans worked it out so far ahead of all the sophisticated people who lived in Greece or Beijing or Delhi or Oxford. Now when I first read about the Mayan achievement, I thought that they had sorted it all out, that they knew how to use 0 as a number. So I was more than a little disappointed when I realised that they only used a shell shape for 0 as a number just to keep two other numbers apart. Naturally, though, putting a 0 between 4 and 5 does make a big difference to your pay at the end of the week.

As we will see in the next section, they could also have found the difference between two numbers. Their numeral system was constructed using dots and sticks, along with their zero. With this notation Mayans could cancel out the sticks or dots when subtracting. However, I'm not sure if they actually did this in practice.

So how and when did the rest of the world finally catch up? Well, it took a long time and a great deal of effort as well as a few bad attempts even after they got the idea of the symbol 0. But first what held them, and so us, up? Well, it turns out that you have to have a place value system, something like we have today, to even worry about place holding. (And the Mayans had this.) For instance, in the previous section we looked at the Roman numerals that we still see. Even this book is infested with them, see the page numbers in the table of contents, for example. But if Claudius had 5 children and the Visigoths abducted them he didn't think of writing down $V - V = 0$. All the problems that the Romans and the Greeks considered involved real things, not that Claudius' children weren't real, but zero was a silly answer and negative answers weren't real. Any measuring problem needed a positive answer. Basically, they lived in a positive world, mathematically speaking.

Who then decided to have place valued systems of numerals apart from the Mayans? Well, it so happens that thousands of years ago the Babylonians did. But they didn't have a zero! They got along without it by the context of the arithmetic. If they wanted to measure the width of the house and they wrote down 123 cubits, that's about 2,583 inches, then they would never think of the measurement really being 1,023 cubits. After all, a house of that width had never been built in that area of the countryside. As the years went by, to roughly 1700 BCE, houses got bigger and maybe in a few cases they didn't know whether they were measuring 123 cubits or 1,023 cubits. To fix up the problem, they put two little wedges down where we put a 0. So they got a zero-type numeral.

But by 700 BCE at Kish, a place east of Babylon, they started to use three hook shapes to mark where the "0" should have been, and sometimes they got lazy and only put one hook. The interesting thing about the tablets that held these numbers was that they never used either symbol for a zero in the units. That means that they had 95, which may have meant 95'. Presumably 95' was so much bigger than 95 that there was never any difficulty interpreting the actual length.

Going back to the Greeks, it's worth noting that there was a small class of their society that needed and used place holders. These were the astronomers who recorded astronomical data. You can see that there was likely to be a big disparity between distances in the Solar System if zeros got dropped off, especially if they were the last six digits. Angles had to be precise and the placeholder was needed then too. Consequently the astronomers actually did use 0, or perhaps more likely O, as their placeholder.

As this was the beginning of 0 being used for zero, it's worth spending some time on the reason for using this roundish symbol or letter. There are several suggestions as to why this symbol was used. Perhaps the most interesting is the idea that it followed from games played in the sand. Coincidentally, the games were played with coins, circular coins. When the coins were taken out of the game a round shape was left in the sand. These shapes were worth *nothing*.

Claudius Ptolemy was a Greek astronomer who lived in Alexandria (like Euclid). In 130 CE he used Babylonian numerals and did insert a 0, not as a number but as a placeholder. As elsewhere, the 0 was only used by astronomers. But they seemed to get tired of it and after a while it was no longer used. A big opportunity had been missed.

The first major step forward was in India and it was from this that our current system was to evolve. Before 500 CE, Indian mathematics used a positional system of numerals with a name for zero. However, the first known inscription that used a symbol 0 as a placeholder was on a stone tablet in 876 CE. We quote from *A History of Zero*, MacTutor, again

> The inscription concerns the town of Gwalior, 400 km south of Delhi, where they planted a garden 187 by 270 hastas which would produce enough flowers to allow 50 garlands per day to be given to the local temple. Both of the numbers 270 and 50 are denoted almost as they appear today although the 0 is smaller and slightly raised.

The placeholder problem seems to have been settled, but what about 0 as a number? We'll get to that, but first what is the problem about having a "nothing" as a number? How do you add, multiply and so on with zero? What are the rules? India again comes to the fore with two mathematicians who tried hard but didn't quite get it right.

Zero in Indian Mathematics

Brahmagupta, whom we will meet again shortly in Sections 8.3 and 8.5, wrote on the subject in about 630 CE. So how did he do arithmetic with zero? First you have to realise that he was trying to treat zero as a normal number. The problem is that it isn't, but he was unable to say no to division by zero. We list his "axioms". Overall not a bad effort, but you might be able to see some difficulties here.

Addition:

1. If zero is added to a negative number the answer is negative;
2. If zero is added to a positive number the answer is positive;
3. If zero is added to zero the answer is zero.

Subtraction:

4. If a negative number is subtracted from zero the answer is positive;
5. If a positive number is subtracted from zero the answer is negative;
6. If zero is subtracted from a negative number the answer is negative;
7. If zero is subtracted from a positive number the answer is positive;
8. If zero is subtracted from zero the answer is zero.

Multiplication and division:

9. If a positive or negative number is multiplied by zero the answer is zero;
10. If a positive or negative number is divided by zero the answer is a fraction with a denominator of zero;
11. If zero is divided by a positive or negative number, the answer is either zero or is a fraction with a numerator of zero and a denominator equal to the number that is dividing;
12. If zero is divided by zero the answer is zero.

Look at these axioms from the point of view of the first person to introduce both negative numbers and zero into arithmetic. Brahmagupta is doing well until he tries to characterise division. I suppose that there is nothing wrong in axiom 10. Writing that dividing n by zero gives $n/0$ isn't a problem – just useless, because you cannot treat $n/0$ as a number. If you divide 23 by 0 to get $23/0$ then, presumably, multiplying $23/0$ by zero gives you 23. But the result of multiplying a number by 0 is 0, by axiom 9.

As for axiom 11, Brahmagupta had a bet each way. He would have done better to have taken the plunge and let the answer be zero. But the axiom 12 is just a step too far. Hmm, why?

The next Indian contributor was Bhaskara II, whom we will also meet again in Section 8.5. His thoughts about 0 were written about 1180 CE. His take on $n/0$ was to use the term infinity for it. Basically, Bhaskara II saw $n/0$ as something very big and unimaginable. This is quite reasonable. Consider $15/0$, but start with $15/r$, where r is a small number. If $r = 1$, you get 15; if $r = 1/2$, you get 30; if $r = 1/3$, you get 45. You can see that the answers are getting bigger. As r approaches zero, $15/r$ gets bigger and bigger and might be said to approach a very big thing like, well, infinity. But now the problem is that if $23/0 = $ infinity, then $23/0 \times 0 = 23 = $ infinity $\times 0$. But $24/0 \times 0 = 24 = $ infinity $\times 0$ too. So $23 = 24$ (!).

However, Bhaskara II did make two steps forward. He realised that

$$0 = 0 \times 0 = 0^2 \quad \text{and} \quad \sqrt{0} = 0,$$

so he was moving in the direction of 0 as a proper number.

Meanwhile, things had been developing in the Muslim world, where al-Khwarizmi in the early ninth century used zero as a placeholder. He had based his work on an Indian volume. But he appears to have known how the Hindu place value system was used. India seems to have been able to manage basic arithmetic using the digits $1, 2, \ldots, 9, 0$, but there was some difficulty coping with 0 and division.

In Morocco, al-Samawal in the 12th century added $0 - n = -n$, and $0 - (-n) = n$. China got the word from India too and used zero in the *Mathematical Treatise in Nine Sections* written by Qin Jiushao in the mid-13th century; see Sections 6.1 and 6.2. In that book, the symbol O is used for zero.

Fibonacci played his part by introducing the new decimal system to Europe, via his personal presence and the *Liber abaci*. But it seems to have been hard to convert merchants. It wasn't until the 16th century that the "new" numeral system was finally accepted by most of Europe. There are a couple of reasons for this. First, the Hindu-Arabic numerals were far more efficient than Roman numerals in written calculation. This was one of the points that Fibonacci was making in the *Liber abaci*. But this brings me to the second point. Merchants generally didn't do their calculations in writing. They used the abacus, as we have seen. These "computers" had been around for thousands of years all over the world. They held the numbers in place value positions. If at any point of a calculation a zero was necessary, there would be no bead in that position. By using the abacus, merchants actually bypassed the need for a zero symbol.

Mayan Civilisation

The Mayans inhabited the Yucatán peninsula (Figure 8.1). Moving from north to south, this area now contains eastern states of Mexico, Belize, Guatemala, Honduras and El Salvador.

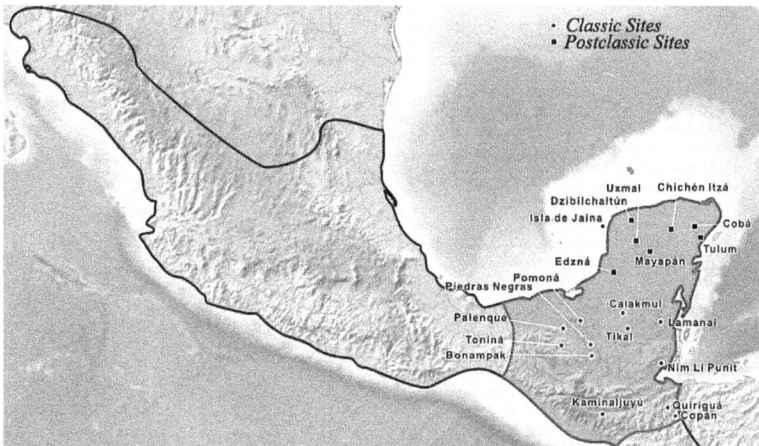

Fig. 8.1   Map of Maya Yucatán (from Wikipedia, Maya civilization)

The Mayan civilisation began in about 2600 BCE, became significant about 250 CE and began to decline around 900 CE. In this latter period, they built large cities, with one of them having a population of the order of 50,000. Cities of this size in Britain, the USA, and Australia are given below, along with Communes of this size in France, for comparison.

Britain: Clacton-on-Sea, Essex, and Kirkaldy, Fife
USA: Newark, Ohio and Jefferson, Indiana
France: La Rouche-sur-Yon, Vendée and Clamart, Hauts-de-Seine
Australia: Bundaberg, QLD and Albury-Wodonga, NSW/Vic border

Apart from their pyramids, perhaps their most outstanding achievement was to develop a numeral system that was well ahead of most (all?) of the world at that time. The system was based on the number 20, the number of digits on the human hands and feet. We show the numbers from 0 to 20 in Figure 8.2. There is a striking resemblance to the Chinese numerals shown in Section 5.2.

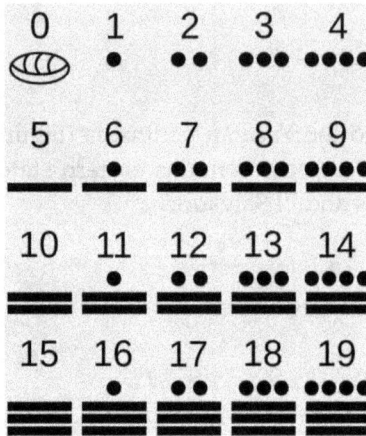

Fig. 8.2   Mayan numerals (from Wikipedia, Maya numerals)

8.3   Brahmagupta

Brahmagupta lived from 598 to 670. We know virtually nothing of his life outside maths and astronomy, even though he was certainly the outstanding mathematician of his age. However, it is probable that he was born in Ujjain, India, in the state of Madya Pradesh. We only know though that he wrote a book called *Brahmasphutasiddhanta* while he was living in Bhillamala (today Bhinmal) because he tells us that in the book itself. In 665, when he was 67, he wrote a second book on astronomy and maths called *Khandakhadyaka*.

We have seen Brahmagupta's contribution to zero and its development in the previous section, but he worked on a number of other things (see Wikipedia, Brahmagupta). For example, he wrote a formula for the solution of a general linear equation which looked like this. If $ax + b = cx + d$, then $x = (d - b)/(a - c)$. He also knew how to solve quadratic equations.

For instance

$$\text{If}\quad ax^2 + bx = c \quad \text{then}\quad x = \frac{\pm\sqrt{4ac + b^2} - b}{2a}.$$

There is not an error here. The standard form in textbooks these days looks at the solution for $a^2 + bx + c = 0$. Hence the change in the term under the square root. Brahmagupta had an equivalent form produced from the one above by dividing the numerator and denominator by 2. He also contributed to the much more subtle problem of finding whole number solutions to quadratic equations; see Section 8.5 on the Pell equation.

Brahmagupta knew the sum of the squares of the first n numbers. Whereas I would write this sum in terms of the number n, Brahmagupta expressed it in terms of n and the *sum* of the first n numbers. This gives

$$1^2 + 2^2 + 3^2 + \cdots + n^2 = \frac{(1+2+3+\cdots+n)(2n+1)}{3}.$$

However, $1+2+3+\cdots+n = n(n+1)/2$ so I can write the answer above in the more usual modern way:

$$1^2 + 2^2 + 3^2 + \cdots + n^2 = \frac{(1+2+3+\cdots+n)(2n+1)}{3} = \frac{n(n+1)(2n+1)}{6}.$$

Brahmagupta also wrote quite a bit on geometry in *Brahmasphutasid-dhanta*. One of these results is called Brahmagupta's formula. This has to do with cyclic quadrilaterals, that is, quadrilaterals whose vertices are on a circle. I've drawn the diagram in Figure 8.3.

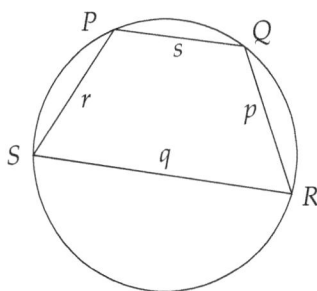

Fig. 8.3   Cyclic quadrilateral

Brahmagupta's formula gives the area of the cyclic quadrilateral, which is

$$A = \sqrt{(v-p)(v-q)(v-r)(v-s)},$$

where p, q, r, s are the sides and v is half the perimeter, $(p + q + r + s)/2$. There is also a Brahmagupta's theorem regarding cyclic quadrilaterals, but it is a little more difficult to explain easily. The formula is of interest because if you make one side of zero length, you have a means of finding the area of a triangle. The triangle version is attributed to Heron of Alexandria, in approximately 60 CE. But it is almost certain that the result was known a couple of centuries before that. A formula equivalent to Heron's can be found in the Chinese book *Mathematical Treatise in Nine Sections*, by Qin Jiushao (1247 CE). His formula is

$$A = \frac{1}{2}\sqrt{p^2q^2 - \frac{1}{2}(p^2 + q^2 - r^2)^2},$$

where p, q and r are the sides of the triangle.

And Brahmagupta had a go at finding a value of π and he got an answer of $\sqrt{10}$. It turns out that this is a pretty good approximation. As a result he managed to find useful things like the volume of a rectangular prism and a pyramid.

Brahmagupta had a strong interest in astronomy. His major achievements were calculating the rising and setting times of heavenly bodies as well as the times of solar and lunar eclipses. Many people at that time thought that the Sun was nearer the Earth than the Moon. Brahmagupta had a solid argument against this based on the illumination of the Moon by the Sun.

Much of Brahmagupta's work was applied to astronomy, so it is not surprising that he was appointed head of the astronomical observatory in Ujjain.

8.4   Al-Khwarizmi

If Brahmagupta is the Hindu in Hindu-Arabic numerals, I suppose that al-Khwarizmi is the Arabic part. His name means that he was a native of Khwarazm, a region near the Aral Sea (see the map of Figure 8.4) that is now partly in Uzbekistan and partly in Turkmenistan. He was born in Khwarazm around 780 CE but appears to have spent most of his life in Baghdad, then Persia, and then Iraq. This was only a (long) camel trip to Bugia where Fibonacci lived a few hundred years later. According to al'Daffa (1978), Muhammad ibn Musa al-Khwarizmi is "In the foremost rank of mathematicians of all time." This is because of his great contributions to mathematics during a lifetime that lasted from about 780 to roughly 870.

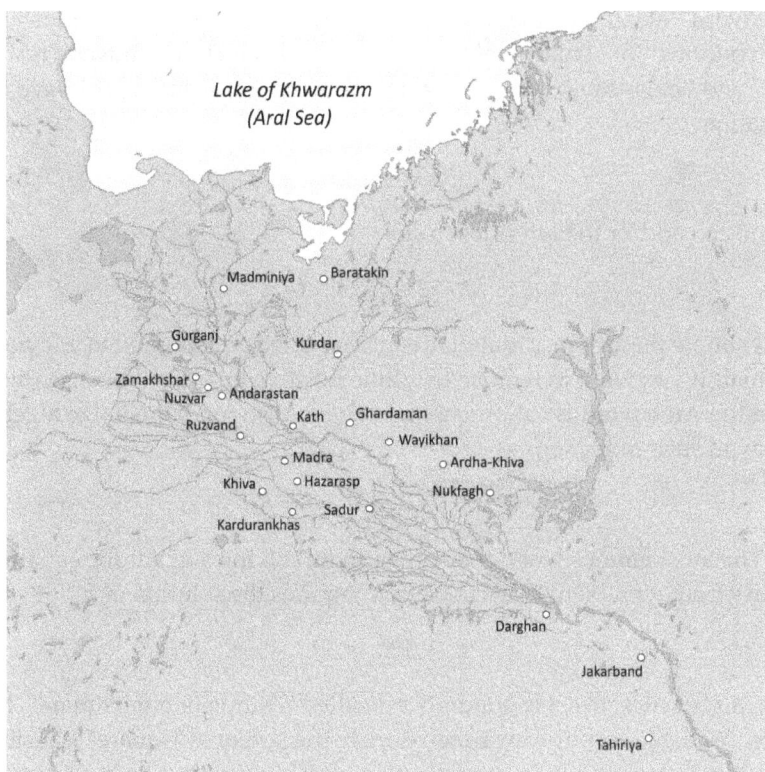

Fig. 8.4   The region where al-Khwarizmi was born (from Wikipedia, Khwarazm)

So what were his main achievements? These were the product of his three major works. Al-Khwarizmi studied algebra for the first time in an algebraic way. Despite not using symbols to represent unknowns he was able to show how to systematically solve linear and quadratic equations. His work, *Hisab al-jabr w'al-muqabala* (in English, *The Compendious Book on Calculation by Completion and Balancing* c. 813–833) was his most important. Al-Khwarizmi describes both the content and audience for his book as

> ...what is easiest and most useful in arithmetic, such as men constantly require in cases of inheritance, legacies, partition, lawsuits, and trade, and in all their dealings with one another, or where the measuring of lands, the digging of canals, geometrical computations, and other objects of various sorts and kinds are concerned.

> Quoted in MacTutor, Biography of al-Khwarizmi

The word "al-jabr" in the title of his book, which is variously translated as "restoring" or "balancing", led to the word "algebra". Indeed "restoring" and "balancing" is basic in algebra today. For instance, if I have the equation

$$5x = 3 - 2x,$$

then I can add $2x$ to both sides to get

$$7x = 3.$$

This simple piece of manipulation can be applied equally well to any polynomial. If you want to remove the same positive quantity from each side, then the Arab word is "al-muqabala". This is also fundamental to algebra today in such reductions as

$$2x = 8 + 5x, \quad \text{hence} \quad 0 = 8 + 3x.$$

The most famous work of al-Khwarizmi, solving a quadratic equation, comes from a problem in his book that requires the solution of

$$x^2 + 10x = 39.$$

The modern way to solve quadratics is called "completing the square", but why? Well, the way al-Khwarizmi does it, the potential "square" is visible, and it is completed quite literally (see also Liu's square root method in Section 5.4).

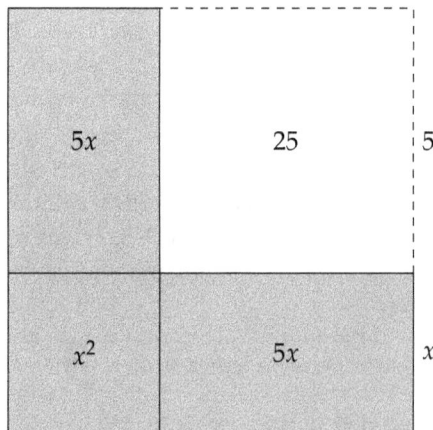

Fig. 8.5   How al-Khwarizmi completed the square

The term x^2 is viewed by al-Khwarizmi as an actual square, of side x. The term $10x$ is a rectangle, with perpendicular sides 10 and x; however he divides it into two rectangles with sides 5 and x, which are placed symmetrically on adjacent sides of the square x^2 (Figure 8.5). This gives a (grey) L-shaped region that can be "completed to a square" in a glaringly obvious way.

Namely, add the white square on the top right, with sides of length 5. This adds $5^2 = 25$ to the L-shaped region, whose area is known to be 39, giving a large square of area 64. Conveniently, a square of area 64 has side 8, so

$$8 = x + 5 = \text{height of L-shaped region,}$$

and therefore $x = 3$. Today, we would be obliged to add that $64 = (-8)^2$, so $x + 5 = -8$ as well. This gives the negative solution, $x = -13$, but al-Khwarizmi was not looking for negative solutions.

Al-Khwarizmi's second important treatise, on astronomy, *Zij al-Sindhind* (Astronomical tables of Siddhanta) was also of great value and contained much original material, but it was based on Brahmagupta's *Brahmasphutasiddhanta*. It contained data on calendars and astronomy, including the motion of the planets, moon and sun. Sine and cosine tables were also included, as were the first tangent tables. The book made a big difference to Islamic astronomy, turning it from a passive study that used other people's research to a research study that they engaged in themselves.

The third book on his list was *Kitab Surat al-Ard* (Book of the Description of the Earth). This was completed in 833. It had a great amount of systematic data on latitude and longitude, locating over 2,000 places of geographical interest.

In addition, a lesser work, translated into English as *Al-Khwarizmi on the Hindu Art of Reckoning*, set out the Hindu numerals and the use of zero as a placeholder. Incidentally, at this time, dust boards were used. The boards were covered in dust or sand and were used to make calculations. It was therefore necessary to have a way of seeing which terms had zero units. This book was full of practical problems that would then have been common in the financial world. It showed a full knowledge of arithmetic in Hindu-Arabic numerals. The Latin version of this book was used in Europe in the 1100s, though it is not clear that this appealed to the merchant classes. A Latin version of the algebra book was translated in 1145 and appears to have had more influence.

The name al-Khwarizmi later led to the word "algorithm" for a set of instructions that solve a particular class of problems. For instance, what we now call the **Euclidean algorithm** (Section 3.2) finds the greatest common divisor of two given numbers. Usually we want an algorithm to produce an answer in a finite amount of time, but in Section 9.3 we will see that it is also interesting when the Euclidean algorithm runs forever.

Getting back to al-Khwarizmi, not everything in his books was original. Fortunately for him, he had access to both Greek works, probably including Euclid's *Elements*, and Indian ones. The Indian material had its roots in a text that was given to the Baghdad court by a political mission from India in 770 CE. So the mathematical baton was passed on again.

8.5 The Pell Equation

The mathematics of Brahmagupta and Bhaskara II also included tremendous progress on the **Pell equation** $x^2 - Ny^2 = 1$, mentioned in Section 3.5 in the case where $N = 2$. There we saw that this case $x^2 - 2y^2 = 1$ has an obvious nonzero solution $x = 3$, $y = 2$, and that infinitely many nonzero solutions can in fact be found, each one obtainable from the one before by a simple algorithm. For small values of N it is not hard to find nonzero solutions of the Pell equation by trial and error. For example, $x^2 - 3y^2 = 1$ has the solution $x = 2$, $y = 1$ and $x^2 - 5y^2 = 1$ has the solution $x = 9$, $y = 4$. But it was not clear whether a solution of $x^2 - Ny^2 = 1$ always exists, for nonsquare N, nor whether one solution would give infinitely many.[1]

The Indian mathematicians Brahmagupta, in the ninth century, and Bhaskara II, in the 12th century, came close to answering both these questions. The second question turned out to be easier and was answered by Brahmagupta by some clever algebra. In modern algebraic notation Brahmagupta's discovery goes like this.

If $x = a_1$, $y = b_1$ is one solution of $x^2 - Ny^2 = 1$, and if $x = a_2$, $y = b_2$ is a second solution (which could be the same as the first!) then another solution is $x = a_3$, $y = b_3$, where

$$a_3 = a_1a_2 + Nb_1b_2, \quad y = a_1b_2 + a_2b_1. \qquad (*)$$

Using this formula repeatedly, we can find larger and larger solutions. For

---

[1] A square N, equal to m^2 say, gives the factorisation $x^2 - Ny^2 = (x - my)(x + my)$, which equals 1 only if the factors are equal to each other, and to ± 1. Equality, $x - my = x + my$, happens only if $m = 0$. This is why we are interested only in nonsquare N.

example, if we start with the solution $a_1 = 3$, $b_1 = 2$ of $x^2 - 2y^2 = 1$, and also count this as the second solution $a_2 = 3$, $b_2 = 2$, then (*) gives the third solution

$$a_3 = 3 \times 3 + 2 \times 2 \times 2 = 17, \quad b_3 = 3 \times 2 + 3 \times 2 = 12.$$

Then, combining $a_3 = 17$, $b_3 = 12$ with the solution $a_1 = 3$, $b_1 = 2$ gives

$$a_4 = 3 \times 17 + 2 \times 2 \times 12 = 99, \quad b_4 = 3 \times 12 + 17 \times 2 = 70,$$

which we can combine again with a_1 and b_1 to get an even larger solution, and so on indefinitely. Notice, incidentally, that these newly found solutions are among those already found by the Pythagoreans using their "side and diagonal numbers" and mentioned in Section 3.5.

So Brahmagupta's formula (*) works when $N = 2$. Why does it always work? Well, if $x = a_1$, $y = b_1$ and $x = a_2$, $y = b_2$ are both solutions of $x^2 - Ny^2 = 1$ we have

$$a_1^2 - Nb_1^2 = 1 \quad \text{and} \quad a_2^2 - Nb_2^2 = 1,$$

and therefore

$$\left(a_1^2 - Nb_1^2\right)\left(a_2^2 - Nb_2^2\right) = 1 \times 1 = 1.$$

Now the magical thing is that the left side of this equation can be rewritten:

$$\left(a_1^2 - Nb_1^2\right)\left(a_2^2 - Nb_2^2\right) = (a_1a_2 + Nb_1b_2)^2 - N(a_1b_2 + a_2b_2)^2, \quad (**)$$

and therefore

$$(a_1a_2 + Nb_1b_2)^2 - N(a_1b_2 + a_2b_2)^2 = 1.$$

This says $x = a_1a_2 + Nb_1b_2$, $y = a_1b_2 + a_2b_2$ is a solution of $x^2 - Ny^2 = 1$, as claimed in (*). We can then go on to produce infinitely many solutions of this Pell equation, as we began to do above for $x^2 - 2y^2 = 1$.

The magical equation (**) that makes all this possible is sometimes called **Brahmagupta's identity**. Don't ask me how he thought of it, but once discovered it is easy to check, simply by multiplying out both sides. It is an "identity" because it is true for all values of a_1, a_2, b_1, b_2 and N. This includes fractional values of a_1, a_2, b_1, b_2. These are not the whole number solutions we really want, but Brahmagupta discovered that fractional solutions of a Pell equation can sometimes lead to whole number solutions.

For example, he discovered that $a_1 = 24$, $b_1 = 5/2$ is a fractional solution of $x^2 - 92y^2 = 1$, which, when combined with itself using (*), gives the whole number solution $x = 1152$, $y = 120$. This happens to be the smallest nonzero whole number solution, so it was unlikely to be found by trial and error. Brahmagupta proudly declared that "a person solving this problem within a year is a mathematician". It is interesting that these discoveries were first brought to an English-speaking audience in the book Colebrooke (1817), but they did not become well-known until the late 20th century.

Colebrooke's book can be found online at

https://archive.org/details/1817-henry-colebrooke-algebra-with-arithmetic-and-mensuration-from-the-sanskrit-of-br

The quote above from Brahmagupta may be found on p. 364 of Colebroke. Also in the book is material from Bhaskara II. Bhaskara II had an algorithm for finding a solution of any Pell equation $x^2 - Ny^2 = 1$. This is a surprisingly difficult task, since the smallest solution is not related in any obvious way to N. I will say more about the method of Bhaskara II in Part 4, which tells how the Pell equation first came to the attention of European mathematicians.

I will just mention Bhaskara II's most spectacular discovery, which is on p. 178 of Colebrooke: the smallest nonzero solution of $x^2 - 61y^2 = 1$ is $x = 1766319049$, $y = 226153980$. Thus $N = 61$ is a remarkably hard case of the Pell equation – a fact that did not go unnoticed when the equation resurfaced in the 17th century.

Coming back to Fibonacci, we will see in the next chapter that he proved essentially the Brahmagupta identity (*) for the value $N = -1$. He proved it, as I suggested above, by multiplying out both sides and seeing that they come to the same thing. However, this took him several pages, because he did all the algebra in words! The Hindu-Arabic numerals were great for arithmetic, yes, but algebra did not yet have a comparably simple method of calculation.

Chapter 9

Mathematics of Fibonacci

9.1 Fibonacci Numbers

Rabbits

The *Liber abaci* did not just show how the Hindu-Arabic system worked; it included many problems to illustrate the processes and to show the advantages of calculating in the Hindu-Arabic numerals rather than the Roman system. One of these problems had to do with rabbits. We quote from Chapter 12 of the book:

> A certain man put a pair of rabbits in a place surrounded on all sides by a wall. How many pairs of rabbits can be produced from that pair in a year if it is supposed that every month each pair begets a new pair which from the second month on becomes productive?

The way that this works is shown in Figure 9.1, where rabbits are shown in different colours. There are three types of rabbit. These are:

- a pair of babies that don't produce, yellow;
- a pair of juveniles can breed and produce a pair of babies after their first month, blue; and
- a pair of mature rabbits who produce a new pair of babies every month, blue.

So, every month a baby pair just gets older; a juvenile pair gets older and produces a pair of babies; and a mature pair gets older and produces a pair of babies. In Figure 9.1, we show how the population and its size (in terms of *pairs* of rabbits), develops from the initial pair. If we start the first month with one baby pair, then the sequence in successive months becomes 1, 1, 2, 3, 5, 8, ... Fibonacci omitted the first term of this sequence.

Month   Population

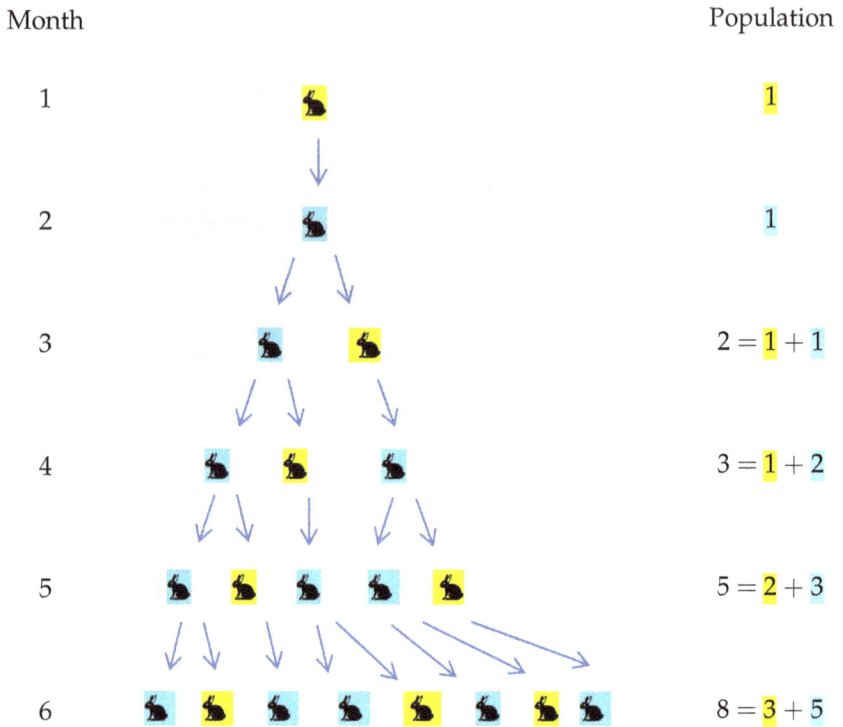

Fig. 9.1   The progression of the rabbit population

You should get some idea from Figure 9.1 how the number of pairs is increasing by looking at the coloured pairs. The yellow new pairs are shown by the smaller numbers in the additions on the right and the blue older pairs contribute to the larger numbers in the additions. What's more the number of blue pairs is always the same as the total number of pairs in the month before. The yellow pairs equal the number of all rabbits two months before. Suppose that we let $F(n)$ be the number at month n. Then the coloured pairs show that

$$F(1) = F(2) = 1;$$
$$F(3) = F(2) + F(1) = 2;$$
$$F(4) = F(3) + F(2) = 3;$$
$$F(5) = F(4) + F(3) = 5;$$
$$F(6) = F(5) + F(4) = 8.$$

It would be worth drawing out the numbers and types of rabbits at the start of seven and eight months. So far, the whole process is summarised in equation (*), which shows that the sum of two consecutive generations of pairs of rabbits is the number of pairs in the next. In other words,

$$F(n+2) = F(n+1) + F(n), \quad \text{where} \quad F(1) = 1 = F(2) \qquad (*)$$

Now the last piece of information, $F(1) = 1 = F(2)$, is extremely important. This is because if $F(1)$ and $F(2)$ had other values, we would get numbers other than 1, 1, 2, 3, 5, 8 for the pairs of rabbits in successive months. The sequence of numbers that have been generated here is called the **Fibonacci sequence** and the terms in the sequence are called **Fibonacci numbers**.

Equation (*) is another example of a **recurrence relation**, seen previously in the work of the Pythagoreans and Liu Hui. It gives the next member of a sequence from previous member(s) of the sequence.

In this chapter we frequently use a recurrence relation that involves the two previous terms. For example, when $f(n+2) = f(n+1) + 2f(n)$ with $f(1) = 1$ and $f(2) = 2$ the first *two* terms give all of the other terms. Thus, $f(4)$ here equals

$$f(3) + 2f(2) = f(2) + 2f(1) + 2f(2) = 2 + 2 + 4 = 8.$$

In this chapter the important recurrence relation is equation (*). Now I'll give a proof of that relation for all n, using algebra.

A Little Bit More Algebra

Let $F(n) = B + J + M$, where B, J, and M are the numbers of pairs of babies, juveniles, and mature rabbits, respectively, at month n. One month later, the babies have become juveniles, the juveniles have become mature, and both the juvenile and mature rabbits have reproduced their own numbers in new baby rabbits. This means that the total number of rabbits is now

$$F(n+1) = (J + M) + B + (J + M),$$

because now $J + M =$ number of baby rabbits, $B =$ number of juveniles, $J + M =$ number of mature rabbits. It follows in turn that, after another month, we get $B + J + M$ new babies, $J + M$ juveniles, and $B + J + M$ mature rabbits. Therefore

$$F(n+2) = (B + J + M) + (J + M) + (B + J + M).$$

To sum up, and simplify, we have

$$F(n) = B + J + M, \quad F(n+1) = B + 2J + 2M, \quad F(n+2) = 2B + 3J + 3M,$$

from which it is clear that $F(n+2) = F(n) + F(n+1)$. So I have proved that the rabbits have the recurrence relation of equation (*), that I said I would.

Fibonacci's point in stating this problem was not to make a true model of rabbit population growth but to show how arithmetically efficient the place-valued Hindu-Arabic system was. Consequently, he has ignored the possibility that some rabbits will die. Or, perhaps he just ignored it because he didn't get such a nice recurrence relation that way. But he did at least keep out stray rabbits by introducing the wall. So we know that stray rabbits don't get into the fluffle[1] of rabbits that we have "modelled" above.

But it isn't only rabbits that seem to generate Fibonacci numbers. Fibonacci numbers can be counted on spirals that occur in pine cones, sunflower heads, and many other things. In addition, Fibonacci numbers also occur in some flowers in another way. Flowers often have a Fibonacci number of petals. For example, Irises have three petals, Buttercups 5, Delphiniums 8, Ragwort 13, and so on. But what does this tell us? Is it just an interesting fact about some parts of nature? Or are we just missing a strong factor in the evolution of certain plants and possibly animals? It should be noted that Hydrangeas have four petals and Frangipanis have six. It might be worth checking if there is a flower for every number of petals up to, say, 12. Anyway, it's clear that Fibonacci doesn't allow his numbers to be used just anywhere. (For a short movie on Fibonacci numbers in nature, see Fibonacci Sequence in Nature, YouTube.) Still, Fibonacci numbers come up all over the place. In the next subsection I've put in a mathematical occurrence that you might find interesting.

Paving a Path

Suppose we have a path that is two units wide and n units long, and we have paving stones that are 2×1 square units in size. To fill the path completely, paving stones can be laid either "crosswise" across the path, or "lengthwise" in parallel pairs. A path of length 1, for example, can be paved in only one way, by a crosswise stone. Figure 9.2 shows all possible ways to pave a path of length 2 or 3.

---

[1] A group of rabbits is called a *fluffle* or a *colony*.

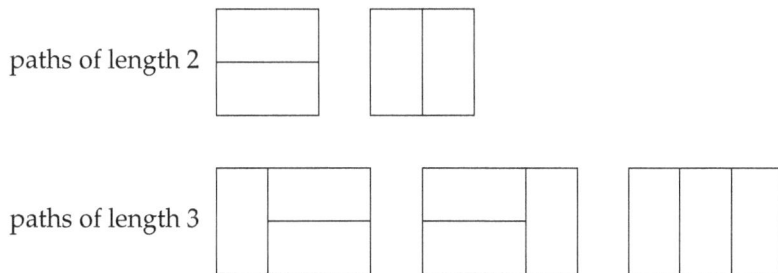

Fig. 9.2   Ways to pave a path of length 2 or 3

So, there are 2 ways to pave a path of length 2, and 3 ways to pave a path of length 3. And we know there is just one way to pave a path of length 1. So far we have the numbers 1, 2, 3 as the number of ways to pave paths of length 1, 2, 3 respectively. Could it be that there are exactly 4 ways to pave a path of length 4? No! The correct answer is 5. You could find this number by drawing all possible pavings, but here is a shortcut.

A paving of length 4 has at its right-hand end either

- two stones lengthwise, in which case the preceding paving has length 2, which can be done in the 2 ways above, or
- one paving stone crosswise, in which case the preceding paving has length 3, which can be done in the 3 ways above.

Altogether, this makes $2 + 3 = 5$ ways to pave a path of length 4.

If you're with me so far, I think you will see that if a path of length n can be paved in $f(n)$ ways, and a path of length $n + 1$ can be paved in $f(n+1)$ ways, then a path of length $n+2$ can be paved in $f(n) + f(n+1)$ ways, namely

- as a paving of length n followed by two stones lengthwise, or
- as a paving of length $n + 1$ followed by one stone crosswise.

Thus the number of ways $f(n)$ of paving a path of length n starts off (for $n = 1, 2, 3, 4$) with consecutive Fibonacci numbers, and it *continues* with Fibonacci numbers, because f satisfies the same recurrence relation:

$$f(n) + f(n+1) = f(n+2), \text{ along with } f(1) = 1 \text{ and } f(2) = 2.$$

9.2 Fibonacci Numbers and the Euclidean Algorithm

When Fibonacci devised his numbers to give his readers practice in adding, he probably did not realise how many mathematical questions lead to them. In later centuries many mathematicians returned to Fibonacci numbers, finding surprising new reasons to study them – and sometimes *old* reasons, because Fibonacci numbers are hiding in mathematics that goes back to Euclid and even the Pythagoreans.

For example, the Fibonacci numbers are a kind of "extreme case" for the Euclidean algorithm. Remember from Section 3.2 that Euclid originally described his algorithm as a process of repeated subtraction, which was later sped up by doing repeated division with remainder instead. However, sometimes the "speedup" is illusory because at every stage b divides a only once, so dividing by b is the same as subtracting b. Here is an example: $\gcd(13,8)$. In this example,

8 divides 13 once, leaving remainder 5,
5 divides 8 once, leaving remainder 3,
3 divides 5 once, leaving remainder 2,
2 divides 3 once, leaving, remainder 1,

and finally 1 divides 2 exactly, so we conclude that $\gcd(13,8) = 1$. We would have obtained exactly the same sequence of numbers simply by *subtracting* the smaller number from the larger, instead of vainly trying to divide. And what is this sequence of numbers? You guessed it – the Fibonacci sequence (without the first term, and backwards)!

More generally, if we start with any pair of consecutive Fibonacci numbers, the Euclidean algorithm trots out all the Fibonacci numbers below them, in descending order down to 1. This is as slow as the Euclidean algorithm gets, yet it is still pretty fast. Each new Fibonacci number is more than 3/2 times the one before, so going backwards through them *reduces* each number to about 2/3 of its size at every step. This is an exponentially fast rate of shrinkage, and the Euclidean algorithm shrinks other numbers even faster. In practical terms, this makes it feasible (using a computer) to find the gcd of numbers with millions of digits.

Thus the Fibonacci numbers answer a question that in some sense goes back to Euclid – how fast is the Euclidean algorithm? – though to formulate the question we need to think of numbers written as Hindu-Arabic numerals, as Fibonacci would want us to do.

9.3 Golden Objects

"Golden" is an adjective that is often used with respect to a certain ratio. I'll talk about that first, and I'll come back to other "golden" objects and their relationship to the Fibonacci numbers.

The Greek letter φ, pronounced phi, is used to denote the ratio AB/AC where the points A, B, and C are such that $AB/AC = AC/BC$, as shown in Figure 9.3. We also say that C divides AB in the *golden ratio*.

Fig. 9.3   Line segments in the golden ratio

Since $AB = AC + BC$ we can rewrite the equation $AB/AC = AC/BC$ as

$$\frac{AC + BC}{AC} = \frac{AC}{BC}, \quad \text{that is,} \quad 1 + \frac{BC}{AC} = \frac{AC}{BC}.$$

Then, by definition of φ, we get

$$1 + \frac{1}{\varphi} = \varphi, \quad \text{or, multiplying through by } \varphi, \quad \varphi + 1 = \varphi^2.$$

This is the quadratic equation $\varphi^2 - \varphi - 1 = 0$, whose solutions are

$$\varphi = \frac{1 \pm \sqrt{5}}{2}.$$

But φ is positive from its definition, so $\varphi = (1 + \sqrt{5})/2$, which is approximately equal to 1.618033988. The other root $(1 - \sqrt{5})/2$, is sometimes labelled ψ (pronounced psi). Both roots are irrational, because $\sqrt{5}$ is, but we will soon see another, more direct, proof that φ is irrational.

The golden ratio appears in several interesting geometric figures. Some of them – a golden triangle, a golden rectangle, and a golden pentagon – are shown in Figure 9.4. You will notice that the "golden" pentagon is simply the *regular* pentagon (equal sides, equal angles). This is just one of many beautifully symmetric geometric objects that involve the golden ratio.

The golden triangle turns out to have an angle of 36°, or $2\pi/10$, at its apex, so 10 golden triangles exactly fill the space around a point, like slices of a pizza, so as to form a regular decagon (10-gon). This creates another regular pentagon, underlining the "fiveishness" of the golden ratio.

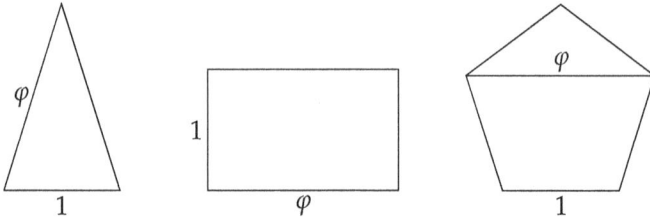

Fig. 9.4　Golden triangle, rectangle, and pentagon

The adjective "golden" has been used because people seem to think that these shapes have some particular beauty. This is especially true of the golden rectangle. It is particularly liked by artists and architects who think that it is in some sense the perfect rectangle. For a long time, it was thought that the Parthenon in Athens had this shape, but that now appears not to be the case. It seems to me that, from a strictly mathematical point of view, the actual golden rectangle in art is somewhat doubtful too. For example, some people claim to find golden rectangles in Leonardo da Vinci's *Vitruvian Man* (Figure 9.5), but the sides for the torso rectangle and the arm and torso rectangle are chosen from different parts of the man's stomach and armpits.

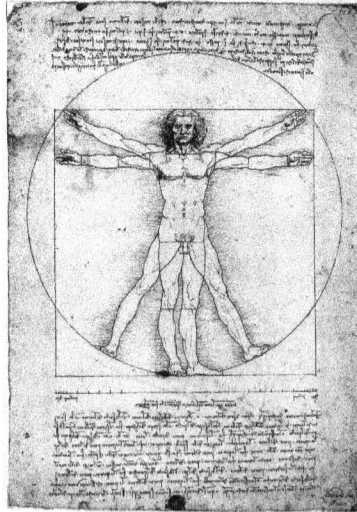

Fig. 9.5　Leonardo's *Vitruvian man* (from Wikipedia, Vitruvian Man)

Returning to mathematics, a spectacular appearance of golden rectangles is in the **icosahedron**, the regular polyhedron with 20 faces mentioned briefly in Section 2.7 as one of the five regular polyhedra. Euclid knew a lot about the icosahedron, but perhaps not this fact, which was pointed out in 1494 by Luca Pacioli, a friend of Leonardo da Vinci. The 12 vertices of an icosahedron lie at the corners of three golden rectangles, which are interlocked as shown in Figure 9.6. It is a fairly straightforward, but lengthy, exercise in Pythagoras' theorem to calculate that $AB = BC = CA$, so that ABC is an equilateral triangle. Altogether, there are 20 such triangles in the figure, so they are the faces of an icosahedron.

The structure of three golden rectangles could in principle be modelled by three postcards, because postcards are similar in shape to golden rectangles, though it is a challenge to get the postcards to interlock properly.

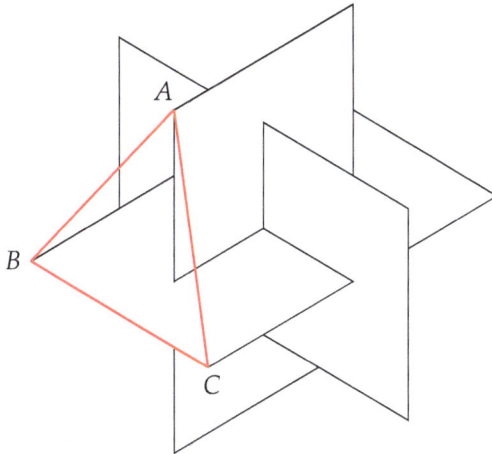

Fig. 9.6   Pacioli's construction of the icosahedron

Irrationality of the Golden Ratio

The golden rectangle has a nice "self-similarity" property that comes directly from the definition of the golden ratio. Namely, if we cut a square off a golden rectangle, as shown by shading the square grey in Figure 9.7, then the white rectangle that remains has the same shape as the original.

The original rectangle has shape determined by sides in the ratio of φ to 1, whereas the white rectangle has sides in the ratio 1 (its "long" side)

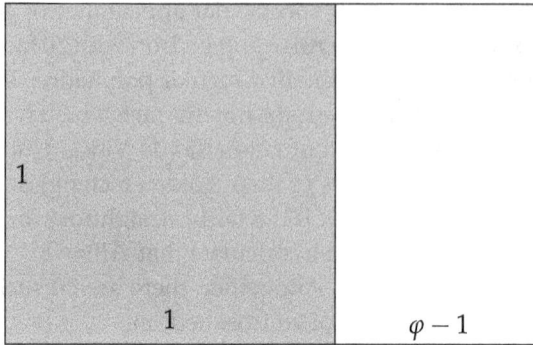

Fig. 9.7   Cutting a square off a golden rectangle

to $\varphi - 1$ (its "short" side). Now

$$\frac{1}{\varphi - 1} = \frac{\varphi + 1}{(\varphi + 1)(\varphi - 1)} \qquad \text{(multiplying by } \varphi + 1\text{)}$$

$$= \frac{\varphi + 1}{\varphi^2 - 1}$$

$$= \frac{\varphi^2}{\varphi} \qquad \text{(because } \varphi^2 = \varphi + 1\text{)}$$

$$= \varphi.$$

So the white rectangle has the same shape as the original rectangle, as claimed.

This means we can similarly cut a square off the white rectangle, leaving yet another rectangle of the same shape, and so on, forever. But cutting off a square amounts to *subtracting* the shorter side from the longer, which is exactly what we do in the Euclidean algorithm – so this appears to be a case where the Euclidean algorithm runs forever! Euclid himself had no qualms about applying his algorithm to possibly irrational lengths, and neither should we.

In fact, by showing the Euclidean algorithm runs forever on the pair of numbers φ and 1, *we have proved that φ is irrational!* Because if φ were a ratio a/b of whole numbers, then applying the Euclidean algorithm to φ and 1 would be just like applying it to the numbers a and b (only scaled down in size by a factor $1/b$). In that case, the algorithm would *not* run forever, so we have a contradiction. Therefore, it is wrong to suppose φ is rational.

9.4 Golden Rectangles and Fibonacci Rectangles

While the Euclidean algorithm cannot run forever on a pair of whole numbers a and b, we know that it runs "as long as possible" on a pair of consecutive Fibonacci numbers – in the sense that division with remainder takes as many steps as repeated subtraction. This suggests we can approximate its behaviour on a golden rectangle by running it on a rectangle whose sides are consecutive Fibonacci numbers. Let's take the example we began with in Section 9.2, 13 and 8, and rerun it by cutting squares off rectangles. We begin with a 13×8 rectangle, cut off an 8×8 square, leaving an 8×5 rectangle, from which we cut off a 5×5 square, and so on. Figure 9.8 shows the result, ending in two 1×1 squares, after following a kind of spiral path.

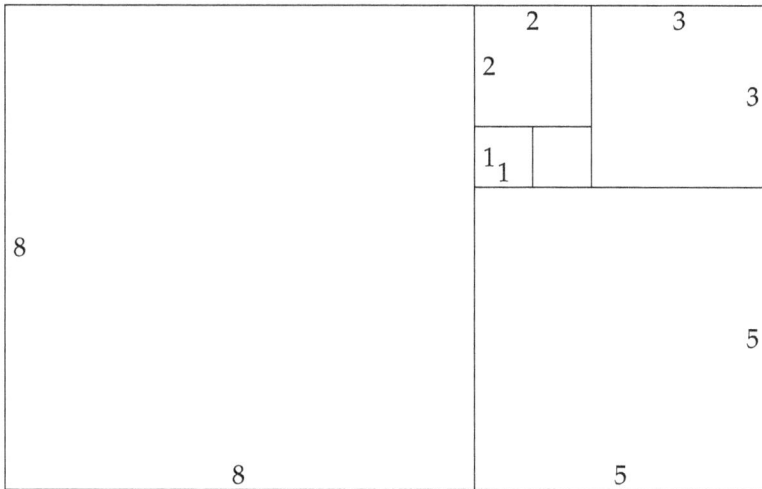

Fig. 9.8 Geometric form of the Euclidean algorithm on 13 and 8

If we begin with a larger pair of consecutive Fibonacci numbers then the process of cutting off squares will run for longer, suggesting that the rectangle with sides $F(n + 1)$ and $F(n)$ approaches more and more closely to a golden rectangle. This is indeed the case, and it shows that the ratio $F(n + 1)/F(n)$ of consecutive Fibonacci numbers approaches more and more closely to the golden ratio φ. For example, $377/233 = 1.61802\cdots$, which agrees with $\varphi = 1.61803\cdots$ to four decimal places.

This suggests in turn that the irrational number φ is subtly related to the Fibonacci numbers, which perhaps explains why there is no obvious formula for $F(n)$. In fact, no formula for $F(n)$ was discovered until the 1720s, when Daniel Bernoulli and Abraham de Moivre independently discovered the formula

$$F(n) = \frac{\varphi^n - \psi^n}{\sqrt{5}} = \frac{\varphi^n - \psi^n}{\varphi - \psi} = \frac{(1 + \sqrt{5})^n - (1 - \sqrt{5})^n}{2^n \sqrt{5}}.$$

Also, since

$$\varphi\psi = \frac{-1 + \sqrt{5}}{2} \times \frac{-1 - \sqrt{5}}{2} = \frac{1 - 5}{4} = -1,$$

we can replace ψ by $-1/\varphi$, and hence write a formula for $F(n)$ in terms of φ alone.

From the formula of Bernoulli and de Moivre we can confirm that the ratio $F(n + 1)/F(n)$ approaches φ as n increases, because

$$\frac{F(n + 1)}{F(n)} = \frac{\varphi^{n+1} - \psi^{n+1}}{\varphi^n - \psi^n}$$

$$= \frac{\varphi - \psi^{n+1}\varphi^{-n}}{1 - \psi^n\varphi^{-n}}. \qquad \text{(dividing by } \varphi^n\text{)}$$

This expression approaches $\varphi/1 = \varphi$ as n increases, since both ψ and φ^{-1} have absolute value around 0.618, so their powers rapidly approach zero. The formula also explains the exponential growth of the Fibonacci numbers, since it shows that $F(n) \approx \varphi^n/\sqrt{5}$.

To see how good this approximation is, let's look at $\varphi^{10}/\sqrt{5}$. This equals $55.003\cdots$, so it agrees with the exact value 55 of $F(10)$ to two decimal places (and to four significant figures). In fact, even for smaller values of n, $F(n)$ is always the nearest whole number to $\varphi^n/\sqrt{5}$.

The "Fibonacci Spiral"

The somewhat spiral path I followed in cutting squares off the downward sequence of Fibonacci rectangles above is sometimes prettified as a sequence of quarter circles inside the successive squares, as shown in Figure 9.9. A quarter circle of radius 8 is smoothly joined to a circle of radius 5, which is smoothly joined to a quarter circle of radius 3, and so on. You can see how to extend this spiral *outwards*, too, by tacking on a square of side 13 at the top, then a square of side 21 to the right, and so on.

To the untrained eye, this looks rather like the curve of a nautilus shell (Figure 9.10). However, that nautilus curve is better modelled by the **equiangular spiral**, also known as the **logarithmic spiral**.

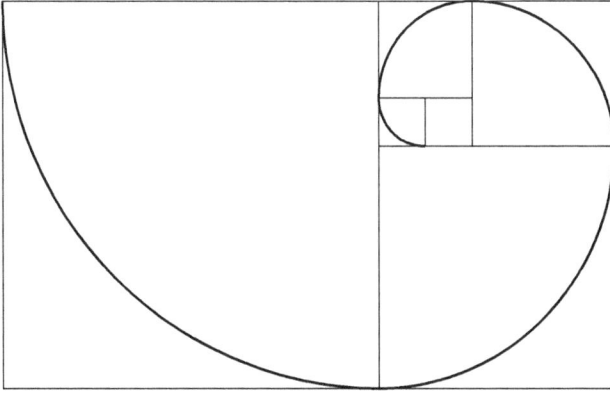

Fig. 9.9   The Fibonacci Spiral

Fig. 9.10   The nautilus logarithmic spiral (from Wikipedia, Nautilus)

In polar coordinates an equiangular spiral has the equation $r = e^{k\theta}$, where r denotes the distance from the origin to the curve in direction θ, and k is some constant.

The equiangular spiral has a "centre" point, the origin, from which each line to the curve makes the same angle (not a right angle). There is no such point for the Fibonacci spiral – instead there is a different "centre" for each quarter circle, and this "centre" jumps along a "square spiral" path.

9.5   Fibonacci's Number Theory

Egyptian Fractions

An interesting novelty in the *Liber abaci* involves the so-called **Egyptian fractions**, an ancient approach to fractions that is not very practical but poses some interesting problems. Fractions were slow to emerge in ancient mathematics, and probably the first fractions to be recognised as numbers were simple parts of a whole such as 1/2 and 1/3. These are examples of **unit fractions**, which we would define as fractions with numerator 1. Other proper fractions, such as 5/6, can be written as sums of distinct unit fractions; in this case

$$\frac{5}{6} = \frac{1}{2} + \frac{1}{3}.$$

For reasons that are not clear, at least to me, the ancient Egyptians hit on the idea that any part of the whole might be represented as a sum of *distinct* unit fractions.

Take 5/7, for example. This fraction is greater than 1/2, so we might start by trying

$$\frac{5}{7} = \frac{1}{2} + \text{something}.$$

This means that our

$$\text{something} = \frac{5}{7} - \frac{1}{2} = \frac{10 - 7}{14} = \frac{3}{14}.$$

Luckily, we can easily spot that

$$\frac{3}{14} = \frac{2}{14} + \frac{1}{14} = \frac{1}{7} + \frac{1}{14}.$$

So in the end we get 5/7 as a sum of three distinct unit fractions:

$$\frac{5}{7} = \frac{1}{2} + \frac{1}{7} + \frac{1}{14}.$$

In this example we were definitely lucky at the end, but the first step made good sense: pick the *largest* unit fraction that is less than or equal to the fraction you are trying to represent. This "greedy" strategy, if followed at every step, actually works, as Fibonacci demonstrated in Chapter 7 of the *Liber abaci*.

To prove this fact, consider a fraction a/b, with $a < b$, and suppose that $1/n$ is the largest unit fraction less than or equal to a/b. This means that

$$\frac{1}{n-1} > \frac{a}{b}, \text{ so } b > a(n-1) = an - a \text{ and therefore } a > an - b. \quad (*)$$

Now let's see what we get when we subtract $1/n$ from a/b:

$$\frac{a}{b} - \frac{1}{n} = \frac{an - b}{bn}.$$

This new fraction has numerator $a' = an - b$, so (*) tells us $a' < a$.

Thus the greedy strategy decreases the numerator at every step, and hence the process of choosing the largest possible unit fraction (which necessarily chooses a *different* fraction at each step) eventually ends, by "infinite descent". At the end our fraction a/b is represented as a sum of distinct unit fractions. Thus the Egyptian way of representing fractions really works!

But although the greedy strategy always gives an Egyptian fraction, it does not always give the simplest one. For example, if we insist on using it with 5/7 we find that 1/6 is the largest unit fraction less than 3/14, not 1/7, so at the next step we are looking at

$$\frac{3}{14} - \frac{1}{6} = \frac{9 - 7}{42} = \frac{2}{42} = \frac{1}{21},$$

which gives the unit fraction representation of 5/7:

$$\frac{5}{7} = \frac{1}{2} + \frac{1}{6} + \frac{1}{21}.$$

This example hints at some of the mysterious complexity of Egyptian fractions. There are many questions about them we do not yet know how to answer. For example, can each fraction of the form $4/n$ be represented as a sum of no more than three distinct unit fractions?

Sums of Squares

Sums of squares have been part of mathematics since the discovery of Pythagoras' theorem and Pythagorean triples, which showed us that the sum of squares can itself be a square. However, not every number is a sum of two (whole number) squares; for example the number 3 isn't. The first mathematician to explore the question of *which* numbers are sums of two squares was Diophantus, whom we met in Section 4.3. In typical fashion, Diophantus does not state theorems about sums of squares. Instead, he gives an example making it clear that he understands what is going on.

In the case of sums of squares, the revealing example is mentioned by Diophantus as part of Problem 19 in Book III of his *Arithmetica* (taken from the English translation of the *Arithmetica* by Heath (1897), p. 167):

65 is "naturally" divided into two squares in two ways, namely into $7^2 + 4^2$ and $8^2 + 1^2$, which is due to the fact that 65 is the product of 13 and 5, each of which is the sum of two squares.

Saying "due to the fact" is the clue here: Diophantus apparently knew that the product of a sum of squares by a sum of squares is itself a sum of squares *and* in two different ways. The most plausible explanation for this knowledge is easy to state in the algebraic language of today:

$$(a_1^2 + b_1^2)(a_2^2 + b_2^2) = (a_1 a_2 \mp b_1 b_2)^2 + (b_1 a_2 \pm a_1 b_2)^2.$$

(See what you get with $a_1 = 3$, $b_1 = 2$, $a_2 = 2$, $b_2 = 1$.) You might think this would be hard to prove *without* algebraic language, but that would be to underestimate the mathematical skills of ancient mathematicians.

Even in Fibonacci's time, there was no good symbolism for expressing this fact about sums of squares. Nevertheless, Fibonacci stated and proved it *all in words* in his *Liber quadratorum* (Book of squares) in 1225. You can read the proof, if you have enough patience, on pages 23–28 of the English translation of the *Liber quadratorum* by Sigler in Fibonacci (1987).

If we choose the upper member of each of the pairs $\mp$ and $\pm$ we get

$$(a_1^2 + b_1^2)(a_2^2 + b_2^2) = (a_1 a_2 - b_1 b_2)^2 + (b_1 a_2 + a_1 b_2)^2,$$

which is the $N = -1$ case of Brahmagupta's identity mentioned in Section 8.5. However, Brahmagupta was interested only in positive values of N, because of the connection with the Pell equation, and in any case Fibonacci's proof is probably independent. Today, the proof is only a few lines: we expand both sides and see that they are equal.

$$(a_1^2 + b_1^2)(a_2^2 + b_2^2) = a_1^2 a_2^2 + a_1^2 b_2^2 + b_1^2 a_2^2 + b_1^2 b_2^2 \quad \text{(expanding)}$$

$$(a_1 a_2 - b_1 b_2)^2 + (b_1 a_2 + a_1 b_2)^2 = a_1^2 a_2^2 + b_1^2 b_2^2 - 2a_1 a_2 b_1 b_2$$
$$+ b_1^2 a_2^2 + a_1^2 b_2^2 + 2a_1 a_2 b_1 b_2 \quad \text{(expanding)}$$
$$= a_1^2 a_2^2 + a_1^2 b_2^2 + b_1^2 a_2^2 + b_1^2 b_2^2 \quad \text{(cancelling)}$$

This just goes to show that algebra needs an efficient symbolic language, to go with the language of Hindu-Arabic numerals that Fibonacci introduced for arithmetic. Europe had to wait a few centuries for this development, but when it came there was an unprecedented explosion of mathematical activity – as we will see in the next chapter.

Given that the product of sums of two squares is itself a sum of two squares, Diophantus realised that knowing which numbers are sums of two squares amounts to knowing which *prime* numbers are sums of two squares, since every number is a product of primes. If you look at the primes less than 100, and find the sums of squares among them, you may see what the sums of squares have in common.

Prime	Sum of squares?
2	$1^1 + 1^2$
3	
5	$2^2 + 1^2$
7	
11	
13	$3^2 + 2^2$
17	$4^2 + 1^2$
19	
23	
29	$5^2 + 2^2$
31	
37	$6^2 + 1^2$
41	$5^2 + 4^2$
43	
47	
51	
53	$7^2 + 2^2$
59	
61	$6^2 + 5^2$
67	
71	
73	$8^2 + 3^2$
79	
83	
89	$8^2 + 5^2$
97	$9^2 + 4^2$

First, 2 is obviously a special case, because all the other primes are odd. So we are really trying to find which kind of *odd* primes are sums of two squares.

Well, the odd numbers

$$1, 3, 5, 7, 9, 11, 13, 15, 17, 19, \ldots$$

alternate between

$$1, 5, 9, 13, 17, \ldots \text{ (of the form } 4n + 1)$$

and

$$3, 7, 11, 15, 19, \ldots \text{ (of the form } 4n + 3)$$

With this clue, I'm sure you can see the answer: *the odd primes that are sums of squares are those of the form* $4n + 1$. Diophantus surely cannot have failed to notice this, but it is highly unlikely that he was able to prove it. The first to claim a proof was Fermat, in the 17th century.

Fermat is the subject of our next chapter, and sums of squares will come up again in Section 13.8.

9.6   Zero: Where to from Here?

The purpose of this section is to introduce some new ideas arising from the role of zero in arithmetic, particularly the mathematical structures called **fields**. At this point the emphasis is on "introduction", though I do sketch an application to a claim of Fibonacci about cubic equations and square roots. Once the foundations have been laid here, I'll be able to show more applications later.

Fields depend on the existence of zero, not merely as a symbol used for writing numerals, but as an object that enlarges the scope of calculation. In a nutshell, fields capture the notions of addition, subtraction, multiplication, and division – as we understand them in arithmetic. In particular, fields are subject to exactly the rules of calculation that we use (often unconsciously) when doing arithmetic.

Although these rules were used unconsciously from perhaps the 17th century, surprisingly there was never a Eureka moment when someone realised that it wasn't possible to divide by zero. On the other hand there was a general acceptance, at least by the middle of the 19th century, that this was the case, along with the other rules of arithmetic.

Preliminaries: Inverses and Identities

As the result of inviting zero into the number family we have a number that can count anything you don't have. I don't have a Lamborghini, so I have 0 Lamborghinis. Another underwhelming thing that 0 can do is "nothing". When you add 0 to anything, there is no change. Still, no other number can do "nothing" in this way, so we honour 0 with the title, the **additive identity**. And these nonevents are compensated by facts like

$$5 + (-5) = 0.$$

In other words zero enables every number, even zero itself, to be nullified, or cancelled out, by adding its negative:

$$7 + (-7) = 0 = 378 + (-378) = 0 + (-0) = 0 + 0 = 0, \text{ and so on.}$$

For any number n, there exists a $-n$, such that $n + (-n) = (-n) + n = 0$. I'll call $-n$ the **additive inverse** of n.

"And that has some merit?" I hear you ask. Thank you for the question. You've probably used it already many times. Suppose I wanted to solve the equation $x + 3 = 10$, where x is some number. I know that there is a -3 such that $3 + (-3) = 0$, so we can add the inverse of 3 to both sides of the original equation to get $x + 3 = 10$, then $x + 3 + (-3) = 10 + (-3)$, and so $x = 7$.

The same works for things like $x - 3 = 10$. Here I can add 3 to both sides confident that -3 will be cancelled out by 3 to leave us with x on the left-hand side and give us $10 + 3 = 13$ on the right-hand side. Ergo, $x = 13$ here.

Immediately we have a method of solving some linear equations. But what about $2x + 3 = 10$? For a start we can add the inverse of 3 to both sides of the equation and we get $2x = 7$. Now it would be nice to get rid of the 2 that is multiplying the x. Well, there are **multiplicative inverses** as well as the additive inverses that we've already defined. Clearly

$$2 \times 1/2 = 1 = 3 \times 1/3 = 3/5 \times 5/3$$

and so on. For n, $1/n$ (which we also write as n^{-1}) is the multiplicative inverse because $n \times n^{-1} = 1$.

However, this is where we have to be careful. Remember this is one of the problems faced by the many mathematicians who tried to introduce zero as a number like every other number: zero doesn't have a multiplicative inverse. There is no number plong, such that zero $\times$ plong $= 1$. So, when I introduce multiplicative inverses, I have to say each number n, *not equal to zero*, has a multiplicative inverse.

In that one step we now know how to solve every linear equation in one unknown. The unknown is often called x, but it can be anything you want to use.

Finally, you will notice that there is a **multiplicative identity**, which plays the "do nothing" role for multiplication that 0 plays for addition; namely, the number 1. Like the additive identity, the multiplicative identity is the result of "cancelling" a number by its inverse, though of course the exceptional number zero cannot be "cancelled" by multiplication – zero is *indestructible* by multiplication.

Rules of Calculation

Once the properties of zero are settled, we can make a short list of all the rules for doing calculations with addition, subtraction, multiplication, and division. Writing $a \times b$ as ab to save space, they are:

$a + b = b + a$	$ab = ba$	(commutative laws)
$a + (b + c) = (a + b) + c$	$a(bc) = (ab)c$	(associative laws)
$a + 0 = a$	$a1 = a$	(identity laws)
$a + (-a) = 0$	$aa^{-1} = 1$ for $a \neq 0$	(inverse laws)
$a(b + c) = ab + ac$		(distributive law)

Together, these laws encapsulate all the rules of calculation, normally used unconsciously when calculating with numbers. In particular

- The commutative laws say that *order* of items does not matter when adding them or when multiplying them.
- The associative laws say that the *grouping* of items does not matter when adding them, or when multiplying them. For example, to add a, b, and c we may first add a and b, then add c to the result, *or* we may first add b and c, then add the result to a. Because of this, we can write sums without brackets; for example $a + (b + c)$ as $a + b + c$.
- The identity and inverse laws say how 0, 1 work, with the sign-switching operation $-$ and the reciprocal operation $^{-1}$ creating additive inverses and multiplicative inverses respectively.
- The distributive law, in combination with other laws, allows us to *expand* a product of sums into a sum of products.

Here is a typical expansion, showing where the various laws are used. This may help you appreciate how much is normally done unconsciously in algebraic calculations – and how painful it must have been for Fibonacci to do this kind of calculation in words.

$$
\begin{aligned}
(a + b)(c + d) &= (a + b)c + (a + b)d && \text{(by distributive law)} \\
&= c(a + b) + d(a + b) && \text{(by commutative law)} \\
&= (ca + cb) + (da + db) && \text{(by distributive law)} \\
&= ca + cb + da + db && \text{(by associative law)} \\
&= ac + ad + bc + bd && \text{(by commutative laws)}
\end{aligned}
$$

A collection of objects that satisfies the commutative, associative, identity, inverse, and distributive laws above is called a **field**. Thus these laws can be viewed as the **axioms** defining a field.

Fields of Numbers

However, the theory of fields is beyond the scope of this book. Right now, I only want to mention some fields that help us understand a claim about numbers made by Fibonacci in his book *Flos*. The simplest field is the field $\mathbb{Q}$ of rational numbers – the numbers that can be expressed as fractions. We know that *irrational* numbers, such as $\sqrt{2}$, exist. So a field that includes $\sqrt{2}$ is necessarily bigger than $\mathbb{Q}$. Surprisingly, the best way to study and compare irrational numbers is by looking at fields that contain them.

Problem 2 of *Flos* is to solve the equation $x^3 + 2x^2 + 10x = 20$. As mentioned in Section 7.3, Fibonacci found a solution, accurate to many decimal places. But he also made the provocative remark that the solution number is *not* one of the irrational numbers studied by Euclid in Book X of the *Elements*. Now, it is most unlikely that Fibonacci could have proved this. He was correct, but proofs of this kind of claim did not arrive until the 19th century – and they are best explained using the theory of fields.

Fibonacci's number α is a value of x that makes the cubic polynomial $x^3 + 2x^2 + 10x - 20$ zero. We can say that α has degree 3, since it does not satisfy any equation of lower degree. Fibonacci could have guessed this, though it would have been hard for him to prove. Euclid's numbers, on the other hand, are things like

$$\sqrt{a}, \quad \sqrt{a} + \sqrt{b}, \quad \sqrt{a + \sqrt{b}}, \quad \sqrt{\sqrt{a} + \sqrt{b}},$$

where a and b are natural numbers. If we set x equal to any one of these, and square repeatedly to remove square roots, we get an equation of degree 2, 4, or 8 satisfied by x. Fibonacci could have seen this too, I think, but proving that Euclid's numbers are *not* of degree 3 is another matter. This is exactly where fields can help.

It turns out that to study a number like Fibonacci's α it is enlightening to study the field $\mathbb{Q}(\alpha)$ that α "generates". $\mathbb{Q}(\alpha)$ consists of all the numbers obtainable from α and rational numbers, by addition, subtraction, multiplication, and division. This sounds complicated, but it turns out they are just the numbers of the form

$$p + q\alpha + r\alpha^2, \quad \text{where } p, q, r \text{ are rational.}$$

No higher powers of α are needed, because $\alpha^3 = -2\alpha^2 - 10\alpha + 20$ by the equation that defines α, and the powers $\alpha^4, \alpha^5, \ldots$ can similarly be replaced by combinations of lower powers. Thus each member of $\mathbb{Q}(\alpha)$ is given by three rational "coordinates" p, q, r. So the degree number 3 is also the number of rational "coordinates", or the **dimension**, of the field.

On the other hand, we find that the dimensions of the fields generated by Euclid's numbers are 2, 4, or 8. In this way, using dimension rather than degree, we find that $\mathbb{Q}(\alpha)$ cannot be one of these fields, and therefore Fibonacci's number α is *not* one of Euclid's numbers.

PART 4

Fermat

Chapter 10

Life and Times of Fermat

Fig. 10.1   Pierre de Fermat (from Wikipedia, Pierre de Fermat)

10.1   Background

Pierre's Birth

On 17 August 1601, Piere Fermat was born. This is verified by more than one website. Pierre de Fermat died on 12 January 1665. Fermat's eldest son, Samuel, produced an epitaph for his father then, in which he wrote words to the effect that on the date above, Fermat died at age 57. See the

2016 article by Friedrich Katscher at

```
https://old.maa.org/press/periodicals/convergence/
when-was-pierre-de-fermat-born
```

Now that is a little disturbing because $1665 - 1601 = 57$ is incorrect. Wasn't Samuel worried by this? It's not even 58, which would have been an admissible answer given the context. It is possible though, that Fermat was born between 13 January 1607 and 12 January 1608. Klaus Barner took this idea up in Barner (2001). With further investigation, Barner (2007) was able to reduce the time period to 31 October to 6 December, 1607. All that he had to do then was to look up the records of baptism, but they were all missing for the years 1607 to 1611! So what now?

Was there a Piere Fermat who really was born in 1601? It turns out that Dominique Fermat, the father of Pierre, was also the father of Piere, with one "r"! Dominique was married twice! His first wife, Francoise, had a son, Piere, and a daughter, and all three of them died before 1607. So now it is safe to say that Pierre de Fermat was born between 31 October and 6 December, 1607. And the arithmetic year problem has now gone away even if the inaccuracy of the birth day hasn't.

Fermat's Family

So I can start again. On some yet to be determined day between 31 October and 6 December 1607, inclusive, Pierre Fermat was born in Beaumont-de-Lomagne, Gascony, in the south of France. His father Dominique was by then married to Claire de Long and they had five children including Pierre. Dominique was a rich leather merchant who also had local political power, as he served three years as one of four Beaumont-de-Lomagne consuls.

Incidentally, the picture in Figure 10.1 is undoubtedly a real likeness of Fermat. So this is the first person in this book whose likeness we can believe in.

But I need to add something about Claire de Long. She brought *noblesse de robe* to the Fermat family. (See Wikipedia, Nobles of the Robe.) This rank was obtained by having someone in the family who had held a judicial or administrative role in a high state office. The rank could be passed on to heirs, which enabled Fermat to be a better contender when he decided to take a post in the Parlement du Toulouse (see next section). Claire and Dominique had four children that I can trace. These were Pierre, Clément, Marie and Louise, but there may have been another child.

Incidentally there was also a rank of *noblesse d'epée*, who were the nobles of the sword. This was a group that was long established in France. The award was originally given to men of high class who had fought for their king and were given estates for their services.

Pierre Fermat married Louise de Long, who was a cousin of his mother. Louise gave birth to five children. The eldest was Clément-Samuel, also known as Samuel, who ensured his name in history by writing the epitaph for his father and collecting together all of Fermat's correspondence and other works in 1679. See the *Varia Opera Mathematica d. Petri de Fermat* at

`https://archive.org/details/variaoperamathe00ferm/page/n21/mode/2up`

Samuel also inherited his father's position in the Parlement, as did Jean-Francoise, Samuel's son.

The Fermat family were strict Catholics and Fermat's other son, Jean, was archdeacon of Fimarens, but I've been unable to find where that was. Of the three daughters, one married and the other two became nuns.

French Culture

Given his father's wealth, Pierre clearly had good food, the best clothes and the best education. His household would have included servants who did all the routine tasks, such as cleaning the house, cooking, and looking after the grounds and/or growing crops.

For some reason though, wealthy people thought that washing was bad for them. Consequently, they smelled! To combat this, the rich people certainly covered themselves and their clothes with sweet-smelling dried herbs. They even had pots that slowly burned things such as incense that made the whole room more pleasant to live in. However, this didn't get rid of the lice and so on that lived in their clothes. But clothes for the rich were quite beautiful for both women and men. Fur, satin, lace and silk were commonly worn. Unfortunately for women and girls though, even little girls, they wore corsets, tight on their upper body to both straighten their backs and make them look thinner. But ruffs (see Wikipedia, Ruff (clothing)), which must have been a nuisance to wear around their necks for both men and women, disappeared in the early part of the 17th century. Generally, children wore the same clothes as adults, just scaled down.

There are many pictures of men's and women's clothes at sites such as

`https://fashionhistory.fitnyc.edu/category/17th-century/`

With access to North America, new food items entered the diet of the wealthy. Turkeys were one of these. The cassoulet, a slow-cooked stew containing pork, progressed through the addition of haricot beans from the New World.

La Varenne wrote what was probably the first French cookbook and *haute cuisine* was established by him. His book contained two sections, one for meat days and one for fasting days. His recipes led to lighter and smaller meals than had been the norm. Pies especially were reduced to single-serve size.

Travel outside the major centres wasn't easy. The roads were poor and progress was slow. As a result, travel by cart or coach between cities took a long time. Presumably though, most rich men owned horses, so they travelled more quickly. Men were employed to send messages for their masters. Canals existed in some places and weren't subject to the various impediments on the roads.

As I've noted, France was a Roman Catholic country, but it tolerated Protestants between the Edict of Nantes in 1598 and its revocation in 1685. In some places Protestants were given castles to help defend themselves and their legal cases were often presided over by Protestant judges. So in Fermat's lifetime Protestants were generally well-treated.

10.2 Pierre (de) Fermat – an Outline

Fermat and the Law

Fermat's adult achievements show that he is a pretty well all-round over-achiever. If FIFA had existed in the 17th century, Pierre would have played for France and they would have won 5-0 in the final and he would have scored a hat trick. With a result like that, they would probably have been playing England.

But first I want to look at Fermat's education. It's not clear though, what his early education was. Presumably there would have been some early home education and then he would most likely have gone to the local Franciscan school.

It is possible that Fermat went to the University of Toulouse in 1623 (see MacTutor, Fermat). If so, he would have only been 16 or 17 at the time. This might have meant that he was quite bright, but it was not uncommon then for young men of this age to attend university. It is not clear whether he gained a degree there or not. Fermat next spent some time at Bordeaux. This was certainly the period when he began to undertake serious mathematical research. It is likely that he worked with some ex-students of François Viète (see Section 11.2). At any rate, he got interested in and studied Viète's work, some of which was based on ancient Greek authors. From Bordeaux, Fermat moved to Orleans where in 1630 he graduated in civil law at the University.

Like Alexander the Great, Section 1.3, Fermat spoke six languages. Fermat's were French, Latin, Occitan, classical Greek, Italian, and Spanish. Occitan (see Wikipedia, Occitan language) is certainly not a language I had heard of before writing this book. In a range of dialects, it covered the southern region of France and parts of northern Spain, including Catalonia and the Basque region. Occitan is an unusual language in that there is no formal written standard. Attempts to formalise this are hampered as the number of people speaking the language in France now is less than 100,000 and their numbers are declining.

Fermat's graduation from the University of Orleans was the end of his formal education and at this point he returned to Toulouse, where he was able to acquire the offices of a consul at the Parlement de Toulouse. There are a few things that need to be explained here. First, "Parlement" at that time did not mean "parliament", even though the French word became anglicised to mean that. "Parlement" was a place of talking ("parler" means to talk) where legal issues were resolved. There were Parlements in other areas of France too. So it was in the lower chamber of the Parlement that Fermat began his career in civil law. Second, his new legal position was obtained by purchase, so he certainly had access to a significant amount of money at this point. Third, "consul" means a lawyer working at the Parlement. Finally, having now entered practice in law, Fermat was able to call himself Pierre *de* Fermat. Later, in 1638, Fermat was promoted to the upper chamber and in 1653, he rose to the peak of the criminal court.

Promotion in the Parlement was usually based on seniority, so Fermat would not normally have reached this position so soon, despite his talents. However, at this time the plague arrived in Southern France, following the Great Plague of Seville, 1647–1652 (see Wikipedia, Second plague pandemic). Many senior members of the Parlement died as a result, opening the way for younger men to be promoted. Fermat himself contracted the plague in 1652 but was one of the few who survived.

From 1631 until he died, Fermat lived in Toulouse but worked in both Beaumont-de-Lomagne and Castres. He would almost certainly have had a home in both of these places.

Fermat and Mathematics

Because of his knowledge of ancient Greek, Fermat was often asked to help mathematicians with ancient Greek texts. In 1629 Fermat himself produced *Plane loci*, an updated version of Apollonius' Treatise *De locis planis* (see Wikipedia, Apollonius of Perga). This, and the influence of Viète, led Fermat to produce his work on analytic geometry (see Chapter 12). If Fermat had not been so slow to make it generally available, we would now know analytic/Cartesian geometry as Fermat's plane geometry and not Descartes'. (Fermat's manuscript *Ad locos planos et solidos isagoge*, was published after his death in 1679.) So we might have referred to Cartesian coordinates as Fermat's coordinates. I'll discuss this more in Chapter 12. But I should add right now that this was the beginning of the use of algebra in geometry. From this time onwards geometric objects could be changed into an algebraic form and so their manipulation became far easier. For example, to find out where two curves intersect just solve the two algebraic equations involved.

With the translation of geometry into algebra it became possible (in the 19th century) to solve the ancient Greek problems of doubling the cube and trisecting the angle. (See Section 2.5 for the problems and Section 12.3 for an outline of the solution, which resembles the solution of a similar Fibonacci problem mentioned in Section 9.6.)

Fermat was thinking about conic sections by 1629, but it is not certain exactly how long after that he realised that he could use algebra to study them. Everyone seems to agree that he knew by 1636 that conics were described by quadratic equations. It is also known that he was studying Viète's algebra by 1629, so it is not a big stretch to imagine that he put two and two together almost immediately.

A recent and reliable history of algebra, *Taming the Unknown*, by Katz and Parshall (2014) has the following on p. 253 about Fermat, which makes a lot of sense to me:

> ...he spent several years in Bordeaux studying mathematics with former students of Viète, who, in the late 1620s were engaged in editing and publishing their former teacher's work ...

> He attempted to use Viète's formulation in his own project of reconstructing Apollonius's *Plane Loci* ...In fact Fermat believed that Apollonius must have used some form of algebra in developing his theorems ...Fermat thus viewed his task as one of recovering the original analysis that underlay Apollonius's complicated geometric presentation.

So, I still think it is fair to date the idea of algebraic geometry to the late 1620s, regardless of the details. If this is true, Fermat completed his algebraic geometry before Descartes did, except there is Descartes' dream of 1619 to contend with (Section 11.7). Should we then give the prize to the first one with something to show his friends? Fermat had a document that he put away in 1636, but which didn't see daylight until 1679. On the other hand, Descartes had a dream in 1619 and copies of his work for all to see in 1637. You decide.

But Fermat made advances in many areas of maths. These include methods for finding maximum and minimum values as well as producing tangents for certain curves. His work on calculus therefore predated both Newton (Wikipedia, Newton) and Leibniz (Wikipedia, Leibniz). Simmons (2007), p. 98, says that

> Isaac Newton would later write that his own early ideas about calculus came directly from "Fermat's way of drawing tangents."

Fermat and Pascal (of Pascal's Triangle fame) collaborated by mail on developing the foundations of probability. I'll say more about this in the next section. Fermat was able to solve such problems as

> a professional gambler asked Fermat why if he bet on rolling at least one six in four throws of a die he won in the long term, whereas betting on throwing at least one double-six in 24 throws of two dice resulted in his losing.
>
> (see Wikipedia, Fermat.)

It's true to say, too, that Fermat founded modern number theory. His work included the Pell equation, amicable numbers, the method of infinite descent (see Section 3.1), and how to express numbers as sums of two or more squares. But he is best known for Fermat's last theorem, which took over 300 years to be proved (see Wikipedia, Andrew Wiles). I'll take up these ideas and more in Chapter 13.

Fermat also made progress in optics, demonstrating that refraction of light could be explained by assuming that light took the quickest path through given media. This has become known as Fermat's principle of least time, a principle later generalised to other "variational principles" in mathematical physics.

One thing, though, that I'd like to say is that there are not many proofs by Fermat on record. Although he knew ancient Greek work, with its high standards of proof, Fermat seemed reluctant to provide proofs of his own results. This is not uncommon in students I have known, especially the more able maths students. They feel that they know how to do the proof and so there is no need for them to actually write it down. But I also know mathematicians who feel that they have never quite got the proof as well as it could possibly be written. Of course, there is no way of knowing why Fermat rarely wrote proofs. However, written proofs are important. They help the reader to understand a topic and be sure that the supposed theorem is actually true.

What Fermat preferred to do was to write his new work in the guise of problems. This was not unusual at a time when mathematical journals generally didn't exist. When he sent problems that relied on a new theorem he had just discovered, he often left his correspondents frustrated.

10.3 Fermat, Pascal and Probability

As I will say a number of times in this chapter, Fermat was a great writer of letters. Often they were meant to tease the recipients with problems that he had recently solved. But in this section, I want to provide a completely different set of correspondence that went on between himself and Blaise Pascal in 1654, on what is called the **problem of points**. The letters were collaborative, with both of them applying their knowledge to solve a generalisation. Similar interactions between researchers in different parts of the world are still sent but mainly by email and not usually sent to tease. On the other hand, now, colleagues may well turn to Skype or Zoom or other such methods of communication, and there are regular journals in all areas of the subject.

If what I have written below interests you, I suggest that you look at the following articles as they provide the whole story and embellishments. For an introduction that focuses on a simple aspect of the problem you might read

```
https://old.maa.org/sites/default/files/images/upload_
library/46/NCTM/The-Pascal-Fermat-Correspondence.pdf
```

More can be found in Wikipedia, Problem of points. But

```
https://probabilityandstats.wordpress.com/2016/11/06/
the-problem-of-points/
```

is a good reference for a complete story. In addition, it has several examples and appropriate generalisations that cover the topic and is especially worth reading. For an overview of the letters sent between Pascal and Fermat, along with some feel for the history of the problem, see

`https://ebrary.net/118868/history/pascal_fermat_1654`

Finally, you can see the letters themselves, virtually in full and translated into English, at

`https://www.york.ac.uk/depts/maths/histstat/pascal.pdf`

(This file reproduces translations that originally appeared in *A Source Book in Mathematics*, by Smith (1959), and *Games, Gods and Gambling* by David (1998).)

So what is the problem? It's about gambling. There are two gentlemen ...seriously, I'm not sure that gentlewomen or ungentle-women generally gambled in those days. Anyway, two men are playing a game and have decided in advance to play a given number of points. They both put the same bet in the pot. However, for some reason, the game has to be stopped before the given number of points have been played. Assuming that the players have not won an equal number of times at this point, how should the pot be divided between the players? It's important to note here that Pascal and Fermat provided a correct way, in fact two ways, to solve this problem, but they didn't extend the problem to three or more players.

This was not a new problem. It had been discussed in the previous two centuries without any significant progress. It would appear that mathematicians of that time, such as Cardano, Pacioli and Tartaglia, just had difficulty getting their heads around the problem and coming up with the ideas that were needed.

Because the problem had been around so long, it is surprising that Fermat and Pascal knocked it off over a summer of correspondence. Fermat found a method for the problem first, essentially based on **binomial coefficients**. Pascal initially came up with a more complicated solution, even though he understood the binomial coefficients in the context of "Pascal's triangle". They succeeded by taking their eyes off what had passed in the game to focus on what might happen next. This had been the problem with earlier attempts to solve the problem of points. For example, Fra Luca Pacioli in his *Summa de arithmetica, geometria, proportioni et proportionalita* of 1494, divided the stakes in proportion to the number of points that each player had won. This would divide the stakes in the ratio 2:1 in a game where one player has two points and the other player one but also in a game where one player has 100 points and the other player 50. If the games were set up as a win for 101 points, then the stake split is clearly unfair to one of the players in both situations.

Tartaglia, in the middle of the 1500s, recognised the difficulty with Fra Luca's division, and he suggested that the ratio of the difference between the size of the lead and the length of the game. With little effort you might see some problems here too. And Tartaglia felt that the problem had no solution other than litigation!

Cardano seems to have understood that the solution depends on what could happen next, but he failed to count the possibilities correctly.

The key to Fermat's and Pascal's solution was to make a proper count of future possibilities. For example, the same division ought to be made for the following two games: (i) two players were aiming at 7 points, and one had 5 and the other 3 points, respectively; and (ii) two players were aiming at 10 points, and one had 8 and the other 6 points, respectively. In both instances, one player only needed 2 more wins and the other needed 4. See Section 11.5 for the ideas behind the solution.

Fermat was able to be at the right place at the right time for setting up the basis for whole sections of mathematics. One of these was probability theory. His work with Pascal achieved this. What they did is still very important in many areas of commerce.

One thing that I'd like to discuss for a minute is who did the discovery in Fermat's work with Pascal. Certainly Fermat came up with the original solution as he counted all the probabilities that were needed. But Fermat and Pascal were both involved. Pascal wrote down a crucial idea, but how much value did Pascal receive from the questions that Fermat asked and the ideas he came up with? Maybe just having the competition of Fermat to get this problem solved was sufficient to push Pascal to his solution. This kind of interaction is not unique in mathematics or any other field.

10.4   Fermat's Legacy

Fermat's Achievements

There is no doubt that Fermat gave a lot to mathematics, despite his lack of written proofs. His achievements are praised in Bernstein (1996), where he says that Fermat

> was a mathematician of rare power. He was an independent inventor of analytical geometry, he contributed to the early development of calculus, he did research on the weight of the earth, and he worked on light

refraction and optics. In the course of what turned out to be an extended correspondence with Blaise Pascal, he made a significant contribution to the theory of probability. But Fermat's crowning achievement was in the theory of numbers.

Peter L. Bernstein, *Against the Gods*, pp. 61–62

Fermat's Last Theorem (FLT)

Fermat is most famous for writing on a copy of Diophantus' *Arithmetica*:

> I have discovered a truly remarkable proof which this margin is too small to contain.

This was the origin of Fermat's last theorem, or properly Fermat's last conjecture. The conjecture goes back to the discovery of Pythagorean triples, which are positive integer solutions of the equation

$$x^2 + y^2 = z^2.$$

As we know from Section 3.7, there are infinitely many solutions of this equation. But Fermat was unable to find any positive integer solutions of the similar equations $x^3 + y^3 = z^3$, $x^4 + y^4 = z^4$, $x^5 + y^5 = z^5$, ..., and he conjectured (and perhaps thought he could prove) that there are none. For over 350 years, this conjecture nagged away at mathematicians and amateurs and demanded to be settled one way or the other. At a certain institution so many "proofs" were submitted that they had a prepared letter that, after pleasantries, said something like "Your error is on page __". The page number was duly inserted and the letter sent.

It is clear now that Fermat did not have a general solution, but like many other mathematicians after him, he managed to deal with a special case or two (see $n = 4$ in Section 13.6). As the years went by, more and more values of n in the equation

$$x^n + y^n = z^n, \quad \text{with} \quad n > 2$$

were shown to provide no whole number solutions of the equation. But this encouraged others to try to find another n for which there were no solutions, or better still to find in one proof, that there were no solutions for any n bigger than 2. In 1994, Wiles was the first person to achieve this latter result. The solution isn't accessible for most mathematicians, but you can see an outline idea of the solution in Wikipedia, Wiles's proof of Fermat's last theorem.

But was all the hard work worthwhile? In the end was it just a problem that could be solved? If nothing else it stimulated work in abstract algebra. During the 19th century some progress was made by factoring $x^n + y^n$. This approach did not produce a proof for all cases, but it did stimulate work on what became known as **algebraic number theory**.

On the other hand, work in another field led Wiles to a proof of FLT. Wile's proof was actually a side result to a conjecture by Shimura, Taniyama and Weil. This conjecture was about things called elliptic curves, also called curves of genus 1. It had been shown that there was a connection between the conjecture and FLT. Wiles eventually proved the conjecture for a class of examples. Fortunately this proof was sufficient to prove the FLT. So work on one area of maths can certainly lead to developments in another. This supports what I was told in school about maths being all connected in some way.

Analytic Geometry

Then there was analytic geometry, more accurately called **algebraic geometry** today. This, of course, was set up independently and almost at the same time by Fermat and Descartes. What they produced was a revolution in mathematics. It laid the foundations for incredible developments in our subject.

Is it strange that two people independently came up with the same ideas and results at almost the same time in history and in the same country. Did one get the development from the other? Was there espionage involved? There seems to be no basis for this at all. It's not as if the independent, concurrent production of new mathematics of Descartes and Fermat was a unique event. Later, for example, Newton and Leibniz managed to invent/discover calculus at pretty much the same time. There are in fact, many examples of this type of independent and nearly simultaneous discovery.

Most likely this independent discovery happens at times when certain questions arise and, because of recent advances in theory or technique, the time is ripe to answer them. In the case of algebraic geometry it was a combination of Viète's work in algebra and the geometric algebra that had come from ancient Greek times.

Number Theory

Number theory in mathematics, sometimes modestly called "arithmetic", is about the addition and multiplication of whole numbers and integers. This sounds simple, but it includes problems such as finding Pythagorean triples or solutions to the Pell equation, which we know are far from easy. Ancients such as Diophantus worked on problems whose solutions were definitely in number theory. Fermat's work in this area was extensive as he was interested in perfect numbers, amicable numbers, Fermat's primes, and the Pell equation. Of course, FLT was number theory in spades. But there were many other parts of number theory that he worked on including his results on the sums of two squares.

His work in number theory wasn't as fashionable in his day as the topics I've mentioned in this section. Carl Boyer notes this in

http://www.britannica.com/biography/Pierre-de-Fermat:

> The handicap imposed by the awkward notations operated less severely in Fermat's favourite field of study, the theory of numbers, but here, unfortunately, he found no correspondent to share his enthusiasm.

However, in later years his work in number theory provided a basis for significant work around the world. Certainly, Fermat's number theory formed the base of modern number theory. Weil (1984) said

> what we possess of his methods for dealing with curves of genus 1 is remarkably coherent; it is still the foundation for the modern theory of such curves. It naturally falls into two parts; the first one ... may conveniently be termed a method of ascent, in contrast with the descent which is rightly regarded as Fermat's own.
>
> A. Weil, *Number Theory. An Approach Through History*, p. 104

Weil here is talking about a rather technical form of "descent" rather than the general form of descent – equivalent to mathematical induction – used since the time of Euclid to argue that any descending sequence of positive integers is finite.

Weil's book traces the evolution of ideas in number theory after Fermat through the work of several eminent mathematicians. But he also emphasises that, until quite recent times, few mathematicians took number theory seriously. In fact, in the lifetime of the authors, number theory has grown from a fringe specialty – often derided as a "bag of tricks" – to a core discipline with connections to virtually all parts of mathematics. In retrospect, we can see that modern number theory started with Fermat.

Chapter 11

Fermat's Predecessors and Contemporaries

The most important predecessor of Fermat, and his greatest influence, was Diophantus, whom we introduced in Section 4.3. In Chapter 13 we will say much more about the influence of Diophantus on Fermat's number theory. However, other predecessors and contemporaries of Fermat were also influential, and the present chapter is devoted to them.

11.1 Bhaskara II

Bhaskara II, mentioned briefly in connection with zero and the Pell equation in Chapter 8, was an Indian mathematician and astronomer who lived between 1114 and 1185. (The II is to distinguish him from Bhaskara I, who lived about 500 years earlier.) Bhaskara II is also known as Bhaskaracharya, which means Bhaskara the teacher. His most famous book, *Lilivati*, meaning "beautiful" or "playful", is said to be named after his daughter. For detailed information about *Lilivati* see Plofker (2009), pp. 182–191.

He wrote six books on mathematics. The content of three of these is outlined below. Bhaskara II became in charge of the observatory at Ujjain, which is in the Indian state of Madhya Predesh, so he was a leading astronomer in his time as well as a teacher. As mentioned in Section 8.3, Brahmagupta once held the same position.

Later I will talk about Bhaskara II and his mathematics regarding Pell's equation and we have already seen his contribution to the development of zero, in Section 8.2. However, he is famous for other maths and I'll touch on some of this now.

He appears to have been the first to fully accept negative numbers. His notation for them has the number with a dot on top. He knew how to do addition, subtraction, multiplication, and division of negative numbers, and hence was able to solve $x^2 = 9$ using both 3 and -3. He even allowed geometric quantities, such as lengths, to be negative.

Bhaskara II stood on the shoulders of Aryabhata, Brahmagupta and Bhaskara I. One such direct link can be seen in his interest in solving Diophantine equations, especially those of the form $ax = by + c$. There he used a method, called **kuttaka**, which had been invented by Aryabhata somewhere around the turn of the sixth century. (Kuttaka means "pulverise", so it is worth remembering this word in case you ever want to say "pulverise" in Sanskrit.) Aryabhata invented kuttaka, but his description of it was vague. Later Bhaskara I turned it into an algorithm, which is none other than the Euclidean algorithm. Indeed, you can say that the Euclidean algorithm "pulverises" numbers by diminishing them until they can be made no smaller. For a discussion of kuttaka and its application, see Wikipedia, Kuttaka. Bhaskara II used this method to solve problems like:

> The horses belonging to four men are 5, 3, 6 and 8. The camels belonging to the same men are 2, 7, 4 and 1. The mules belonging to them are 8, 2, 1 and 3 and the oxen are 7, 1, 2 and 1. All four men have equal fortunes. Tell me quickly the price of each horse, camel, mule and ox.

I move on quickly because you have already worked out that horses cost 85 units each, camels 76, mules 31, and oxen 4.

Bhaskara II has a book full of these questions. So try this one. It has something to do with arithmetic progressions that I've already talked about in Sections 5.5 and 8.3.

> On an expedition to seize his enemy's elephants, a king marched two yojanas the first day. Say, intelligent calculator, with what increasing rate of daily march did he proceed, since he reached his foe's city, a distance of eighty yojanas, in a week?

In terms of yojanas, the daily increment is the fraction 22/7 that approximates π, though it is surely coincidental that Bhaskara II hit on this number.

Bhaskara II was interested in combinatorial questions such as:

> Consider the number with digits $d_1 d_2 \ldots d_n$, with $d_1 \neq 0$. Take any number N. Then how many n-digit numbers satisfy
>
> $$d_1 + d_2 + \cdots + d_n = N?$$

For example, how many 2-digit numbers are there whose digits sum to 5? The answer here is five because the only possible answers are 50, 41, 32, 23 and 14. Does that make it easy to find the number of 2-digit numbers whose digits add up to any N? How about doing the same thing with 3-digit numbers?

Bhaskara II also looked at trigonometric functions and proved that

$$\sin(A + B) = \sin A \cos B + \cos A \sin B.$$

Once he had done that, it was easy to expand $\sin(A - B)$. This was the first appearance of the sine sum and difference rules in India, though a similar result had been found by Claudius Ptolemy in Alexandria in the second century CE. See Van Brummelen (2009), p. 75.

Indians as early as Aryabhata in 499, believed that the planets revolved around the Sun. Bhaskara II was able to measure the length of an Earth year as 365.2588 days, which is very accurate for the tools he had to work with. He was only a minute out.

Among other things he was able to predict lunar and solar eclipses quite accurately. This was due to his invention of something near to calculus. Bhaskara II knew that the "rate of change" of a function at maximum and minimum points is zero. He also found a result that amounts to Rolle's Theorem (see Wikipedia, Rolle's theorem). What he had was that if $f(a) = 0 = f(b)$, then there is a point between a and b where the gradient of f is zero. This has to assume some smoothness.

From these results we can see that Bhaskara II was an outstanding mathematician. His work was world-leading and far ahead of the maths done in Europe in his time. In fact, the Europeans took a few hundred years to catch up with him. For example, Bhaskara II was able to solve Pell's equation $x^2 - Ny^2 = 1$ for $N = 61$ and 67. Brouncker and Fermat only discovered the essence of Bhaskara II's methods in the middle of the 17th century and even then Bhaskara II's presentation was nicer.

11.2　Francois Viète

Viète (see Wikipedia, François Viète) is often known by the Latin form of his name, Vieta, in older books (due to the fact that, like most scholars at the time, he wrote in Latin). He was born in 1540 in Fontenay-le-Compte, which is now in the region Pays de la Loire. This faces the Bay of Biscay and is to the south-east of Brittany. Viète, like Fermat, was a lawyer, and he became privy councillor under the French kings Henry III and Henry

IV. Viète was certainly involved in important legal work. For example, he was involved in the legal affairs of Mary, Queen of Scots. So Viète was another rich lawyer who did mathematics. For Viète, mathematics was done in periods of leisure. One of these periods lasted four years when he was thought to have been too closely linked with the Protestant cause. This period came to a close when Viète was called on to decipher important letters for the two Henrys. Before he died, Viète wrote an essay on cryptography which made all the encryption methods of the day useless.

Viète wrote a foundation for algebra, putting it on the same footing as geometry. His work on algebra included putting letters for variables that made algebra easier to write and read. Using this he produced the binomial theorem that Pascal and Newton later extended. Viète also saw the link between the coefficients of a polynomial and the sums and products of its roots. He also invented ways to find general solutions for equations of degree up to four, as well as finding approximate solutions of quadratics and cubics. The latter were known by Fibonacci, but Viète didn't have access to those results. Strangely, though, in Viète's algebra he didn't use a notation for multiplication or equality.

Further, Viète found a formula for determining $\sin kA$ in terms of $\sin A$. At about the same time he produced the first infinite product for π (Viète's formula):

$$\frac{2}{\pi} = \frac{\sqrt{2}}{2} \times \frac{\sqrt{2+\sqrt{2}}}{2} \times \frac{\sqrt{2+\sqrt{2+\sqrt{2}}}}{2} \cdots,$$

the nth term of which is $\cos \frac{\pi}{2^{n+1}}$.

In 1596, Viète solved a challenge problem of van Roomen, a Dutch mathematician. The problem was a polynomial equation of degree 45. Viète solved the problem very quickly after recognising that it was a trigonometric problem in disguise. Van Roomen's problem is to solve

$$\begin{aligned}
f(x) = {}& x^{45} - 45x^{43} + 945x^{41} - 12300x^{39} + 111150x^{37} - 740259x^{35} \\
& + 3764565x^{33} - 14945040x^{31} + 46955700x^{29} - 117679100x^{27} \\
& + 236030652x^{25} - 378658800x^{23} + 483841800x^{21} - 488494125x^{19} \\
& + 384942375x^{17} - 232676280x^{15} + 105306075x^{13} - 34512075x^{11} \\
& + 7811375x^{9} - 1138500x^{7} + 95634x^{5} - 3795x^{3} + 45x \\
= {}& \sqrt{\frac{7}{4} - \frac{\sqrt{5}}{4} - \sqrt{\frac{15 - 3\sqrt{5}}{8}}} \approx 0.4158234.
\end{aligned}$$

Viète guessed that $f(x)$ is just $\sin 45A$ in terms of $x = \sin A$, so he used the equations for $\sin 5B$ and $\sin 3B$ in terms of $\sin B$, to obtain $\sin 45A$, and hence found the solutions, all of which lie between -1 and 1.

Figure 11.1 shows the polynomial $f(x)$ as van Roomen wrote it in 1593. It begins after the word "ad" in the fourth line, in the reverse of the above order, with the exponents of x written as circled numbers, without the x. This notation was introduced by Bombelli in an algebra book published in 1572. We will see more of Bombelli's work in Section 12.2. Notice that van Roomen uses $+$ and $-$ signs similar to ours, but he places commas in large numerals every four digits. The illustration and some of the history behind it may be found in Van Assche (2022).

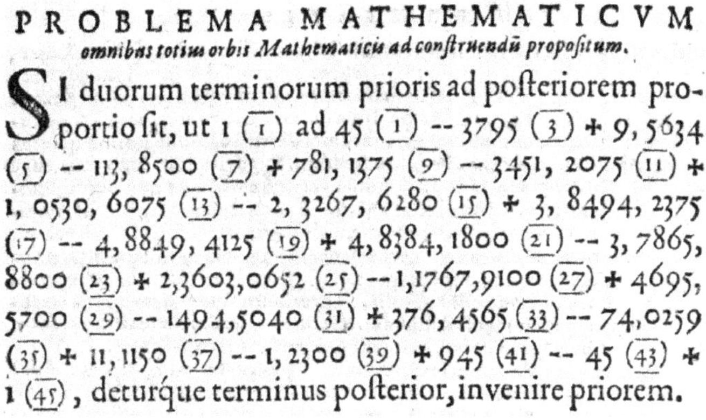

PROBLEMA MATHEMATICVM
omnibus totius orbis Mathematicis ad construendum propositum.

SI duorum terminorum prioris ad posteriorem pro-
portio sit, ut 1 (1) ad 45 (1) -- 3795 (3) + 9,5634
(5) -- 113,8500 (7) + 781,1375 (9) -- 3451, 2075 (11) +
1, 0530, 6075 (13) -- 2, 3267, 6280 (15) + 3, 8494, 2375
(17) -- 4,8849, 4125 (19) + 4,8384,1800 (21) -- 3, 7865,
8800 (23) + 2,3603,0652 (25) --1,1767,9100 (27) + 4695,
5700 (29) -- 1494,5040 (31) + 376,4565 (33) -- 74,0259
(35) + 11,1150 (37) -- 1,2300 (39) +945 (41) -- 45 (43) +
1 (45), deturque terminus posterior, invenire priorem.

Fig. 11.1 The polynomial of van Roomen

Later, van Roomen, wishing to settle a difficulty understanding Viète's solution of a problem, rode his horse from Holland to see Viète. Van Roomen was so impressed by what he saw of Viète's new algebra that they worked together for about a month. Viète paid all van Roomen's expenses and they parted great friends.

Viète died in 1603 in Paris, only a few years before Fermat was born.

11.3 De Méré

Antoine Gombaud, Chevalier de Méré, Sieur de Baussay, lived between 1607 and 1684. As an adult he spent his time between his modest estate in Poitou, France, and at the court in Paris. It's about 400 km from Paris so it would have taken him some time to travel between his home and the court. I'll refer to him as de Méré, but he is also called Gombaud in the literature. For more details of the story I am about to sketch, see Ore (1960).

De Méré had a classical education and then went into the army for a while. At the court of Louis XIV, he quickly became prominent because of his good conversation, good taste and charm. His personality enabled him to handle sticky situations. He also wrote philosophical material and is well-known as a contributor to 17th-century French literature.

As I'll show later though, there were times when he "got too big for his boots". Someone from the court is supposed to have said de Méré thought that "he could teach Madame de Maintenon courtly behaviour and Pascal, mathematics." It is not at all clear that de Méré's teaching enabled Madame de Maintenon to become the king's mistress, or Pascal to become a great mathematician. It is likely that both de Méré and Pascal (before his religious conversion) were gamblers, since many in the court gambled regularly. Yet neither de Méré nor Pascal was a compulsive gambler. In fact, they both wrote against excessive gambling. On the subject of gamblers, de Méré wrote:

> I confess also that on my part I deplore you who are confined to gambling, longing for nothing but luck, without eyes for anything but this artificial world, almost like the courtesans to whom the great beauties of nature are unknown.

While Pascal in later life wrote in his *Pensées*, Part II, Section 139:

> But you may say, what is his object? To boast tomorrow to his friends that he has played better than another? Such a man relieves the tedium of his life by playing every day. If you were to give him, in the morning, the money he might win during the day on condition that he should not gamble, you would make him unhappy. One may think possibly that he seeks the entertainment only of the game and not the gain. But let him play for nothing and he is bored and does not warm up to it. Therefore, it is not only the amusement which he seeks – languishing play without passion he finds tedious. He must become excited and deceive himself into believing that he would be happy to win that which he would not even accept were it not for the play. He must form an object for his passion to excite his desire, his anger, his fear, just like children who are scared of their own faces when they have blackened them.

However, they were both interested in solving problems related to gambling, though Pascal downplayed the financial side and calculated only on the probabilities of winning or losing.

De Méré's Problems

In mathematics, de Méŕe became famous for the problems that he posed to Pascal. There are two of them.

Problem 1: Let a game consist of rolling two dice. How many rolls does a player need to make before his odds of getting two sixes at least once are better than even?

It turns out that this problem had been well-known and solved for centuries before de Méré posed it. Certain aspects of probability were known (for example, to Cardano) before Pascal and Fermat looked at the area.

Pascal's approach, which could be solved by high school students today, went as follows. First, he considered one roll and the probability of not getting a double six. This is $35/36$. The probability of getting no double sixes after two rolls is $(35/36)^2$. For n rolls the answer is $(35/36)^n$. So the probability of getting one double at least is $1 - (35/36)^n$. For $n = 24$, I got 0.491404 and for $n = 25$, I got 0.505522. So, to get at least one double six, your chances are better than even if you roll 25 or more times.

From what we know of Pascal, he published results that he thought were significant. Since he made no attempt to publish this result, he evidently saw no difficulty in it. Later, Fermat and others easily solved this problem too.

De Méŕe suggests that he had solved this by experiment/observation. He too got 25. However, Ore (1960), points out that it would have required a great deal of experimentation to get this result. For a start you can see that the values for 24 and 25 rolls are quite close. To distinguish them, Ore (1960) calculated that a modern analysis would require 100 trials, and that would require thousands of throws. But Ore's clincher is that dice in the 17th century were not as reliable as they are today. The cubes for dice in de Méré's time were made from bone, which would not have been uniformly dense, and so they would have exhibited bias. Good dice could have been made, but only with a great deal of effort.

Problem 2: The question is how one shall divide equitably the prize money in a tournament in case the series for some reason is interrupted before it is completed.

This problem is of course the "problem of points" mentioned in the previous chapter. It was not a problem that de Méré invented. Such problems go back to 1380 at least. In fact, Ore (1960) believes they go back to Arabic times, pre-Fibonacci, where inheritance problems were dealt with.

Two examples of Problem 2 can be found in the 1494 treatise *Summa de arithmetica, geometria, proportioni et proportionalita,* by Fra Luca Pacioli, see

`https://mathshistory.st-andrews.ac.uk/Biographies/Pacioli/`

> A team plays ball such that a total of 60 points is required to win the game, and each inning counts 10 points. The stakes are 10 ducats. By some incident they cannot finish the game and one side has 50 points and the other 20. One wants to know what share of the prize money belongs to each side.

And

> Three compete with the cross bow and the one who first obtains six first places wins; they stake 10 ducats among themselves. When the first has four best hits, the second three, and the third two, they do not want to continue and decide to divide the prize fairly. One asks what the share of each should be.

But Fra Luca went so far as to make some progress, as he seemed to know that players who are behind ask for too much. Incidentally, there seemed to be some reluctance to pose probability problems in a gambling situation. Pascal freed probability problems from gambling too.

There is no evidence that Problem 2 had been solved before Pascal and Fermat tackled it by different methods in 1654. I'll go further with Problem 2 later. Pascal showed his method to the mathematician Roberval,

`https://mathshistory.st-andrews.ac.uk/Biographies/Roberval/`,

but didn't get a pleasant response. This may have been typical of Roberval, because a contemporary is said to have called him

> the greatest mathematician in Paris, and in conversation the most disagreeable man in the world.

Until now I may have given you a very positive view of mathematicians and would-be mathematicians, but perhaps this section has shown that mathematicians are human and have the same range of personalities as any group of people. The rebuff that Pascal got from Roberval may well have motivated Pascal to send his work to Fermat. When Fermat agreed with Pascal's result, Pascal said

> I see that the truth is the same in Toulouse and Paris.

11.4 William Brouncker

Early Life

Brouncker was born around 1620, but the exact date is uncertain (see Wikipedia, William Brouncker). It is also not clear where he was born but it is likely to have been in Ireland as he is known as an "Irish mathematician". His father was Sir William Brouncker and he and his family were closely associated with royalty. This was to become a problem during the Civil Wars that raged in England from 1642 to 1652. Despite this, in 1645 Sir William bought himself a peerage and became the first Viscount of Castle Lyons. Incidentally, Fermat lived through a civil war too, called the Fronde, 1648 to 1653.

Brouncker went to Oxford University at 16, where he studied maths, languages and medicine. It appears that maths at Oxford in those days was not an academic study. It was confined to arithmetic for practical people, such as traders, merchants, carpenters, and so on. But Brouncker did complete a Doctor of Medicine from Oxford in 1647.

Mathematical Achievement

Because of his family contact with royalty, Brouncker kept out of sight in the country during the Civil Wars especially as Oliver Cromwell beat the Royalists and ruled until 1658. However, this period was the most mathematically productive part of Brouncker's life. Brouncker got results on a number of problems. I have presented these in chronological order.

1654. Produced generalised continued fractions for $\pi/4$ and for $4/\pi$. Here I show the latter. (For more on continued fractions see Section 13.5.)

$$\frac{4}{\pi} = 1 + \cfrac{1^2}{2 + \cfrac{3^2}{2 + \cfrac{5^2}{2 + \cfrac{7^2}{2 + \cfrac{9^2}{2 + \ddots}}}}}$$

This continued fraction is not a practical way to produce approximations of π – but what a formula! It was later shown (by Euler in 1748) to be an easy algebraic consequence of another famous formula for π,

$$\frac{\pi}{4} = 1 - \frac{1}{3} + \frac{1}{5} - \frac{1}{7} + \frac{1}{9} - \cdots$$

1654. The quadrature of the hyperbola, showing that the area under the curve $y = 1/(1 + x)$ between 0 and 1 (which is the natural logarithm of 2) equals

$$1 - \frac{1}{2} + \frac{1}{3} - \frac{1}{4} + \frac{1}{5} - \cdots$$

1657. Inspired by a challenge from Fermat, Brouncker found a way to solve Pell's equation $x^2 - By^2 = 1$, probably similar to Bhaskara II's method. Here B is deliberately chosen to stand for both Bhaskara II and Brouncker (and perhaps Brahmagupta). Brouncker solved the equation for several values of B, especially for the case $B = 61$ in one of Fermat's challenges. Fermat would have known the solution for various B before he sent out his challenge letter, because it is surely no coincidence that $B = 61$ is the first really hard case. If not for a slipup by Euler, mistakenly attributing the equation $x^2 - By^2 = 1$ to Pell, we might be calling this equation Brouncker's equation (or Brahmagupta's, or Bhaskara II's).

1659. He found arc lengths of parabolas, cycloids and $y^2 = x^3$. The latter is sometimes called the semicubical parabola, and it was the first algebraic curve to have its arc length determined, by Neil and van Heuraet in 1657. Brouncker improved on Neil's method. The length of the cycloid was first found by none other than Christopher Wren, the architect of St Paul's and many other churches in London. To see what a cycloid looks like see Figure 11.2 below. To see it being generated look at the animation in Wikipedia, Cycloid. There you can clearly see that it is produced as the path of a chosen point on a circle as the circle rolls along a straight line. It's very nice.

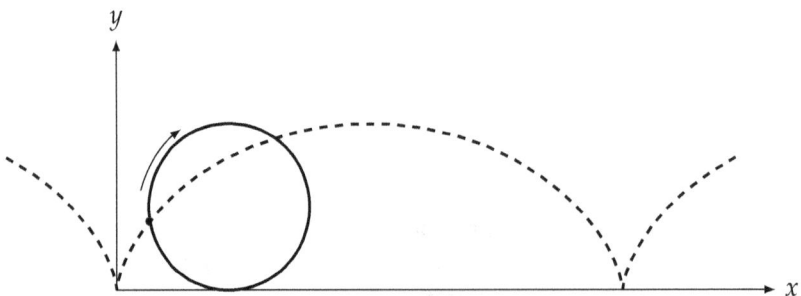

Fig. 11.2   Cycloid as the path of a point on a rolling circle

It is not known who first studied the cycloid. Strangely, the ancient Greeks studied the more complicated *epi*cycloids, obtained by rolling one circle upon another, but seemingly not the simpler case of rolling a circle on a line. The cycloid was first given its name by Galileo. He also spent some time on it and tried to find the area beneath one arch of the cycloid. The way he did it was very hands-on. First he constructed the cycloid on sheet metal. Then he cut out the circle that produced the cycloid and an arch of the cycloid and weighed them. He found that the ratio of the area of the circle to that of the cycloid was 3 : 1, which is the right value. But his next step was bad. For some reason Galileo thought the answer had to be irrational. As such it couldn't be a ratio of whole numbers so he gave up.

Much of Brouncker's work appeared in publications written by John Wallis, an Englishman who was also a correspondent of Fermat. See Wikipedia, John Wallis, and for more information on the mathematics of both Brouncker and Wallis see the thesis of Jacqueline Stedall, which may be downloaded from

https://oro.open.ac.uk/58080/

This thesis was later published as Stedall (2002).

From at least 1657, Brouncker joined a group of scientists who met at Gresham College in London. Their aim was to promote Physico-Mathematical experimental learning. At various times, Brouncker engaged in experiments such as those that involved ballistics and pendulums. This group petitioned the king, then Charles II, to grant them a royal grant of incorporation. This duly occurred and in 1662 the Royal Society of London came into being. Now it is certainly true that Brouncker was a friend of the king, but this is not the only reason why Brouncker became the first President of the Royal Society. He was a wealthy man, and he never married, so it was probably felt that he had more time than others. His ability and availability made him a perfect full-time president. But he got offside with many of the members, Robert Hooke in particular, and in 1677 a motion was moved to elect the Council by vote. Brouncker was very much against this. It is said that "Lord Brouncker in great passion, raved and went out". He didn't attend the meeting when the new Council was elected.

In 1662 Charles II also made Brouncker both Keeper of the Great Seal and Chancellor to Queen Catherine. (See Wikipedia, William Brouncker.)

Brouncker was keen on ships. At one stage he gave the king a boat that he himself had designed and made. Given his interest in ships and the sea, it was perhaps not surprising that in 1664, Brouncker was given a position as Commissioner to the Navy Board. From 1546 to 1832, this body looked after the civil administration of the Royal Navy.

The Great Plague

In 1665 the plague broke out in London. There had been many such events but this was the worst. It is estimated that 100,000 people died out of a population of about 400,000. This was not the last time that London suffered under the bubonic plague, but it was never so rampant as in 1665/6.

The infection was caused by the bacterium Yersinia pestis. It was carried by fleas on rats and small animals and produced flu-like symptoms. In addition, painful swelling occurred near the site of the flea bite. These swollen lymph nodes are called bubos, hence the name bubonic plague. (See Wikipedia, Bubonic plague.) Death generally followed about 10 days after the initial infection. A big percentage of an untreated population will die.

Consequently, the king, the Parliament, and the wealthy fled to the country to be away from London. As Gresham College is in the centre of the city, it is unlikely that there were any scientific meetings there during the period of the plague.

Plagues are, at least in part, exacerbated by dense populations living in bad conditions. You wouldn't have liked the poorer areas of London in the 1660s. Many streets had open sewers in the middle of the roads. Sanitation wasn't valued as chamber pots were emptied from upstairs windows into the street with potentially bad outcomes to either pedestrians or their clothes. The smell was terrible. Similar problems must have affected Fermat during the French plague of 1651.

What stopped the plague? It was once thought that the Great Fire of London that flared up in September of 1666 had done the trick. But this was not the case. It seems the plague just fizzled out. It did reduce the size of the population as people were encouraged to leave the city. That would have helped its decline. Surprisingly, despite some new plans for the fire damaged areas, they were rebuilt much as they had been before, though with many new buildings, such as the famous churches designed by Christopher Wren.

I was living on the outskirts of London in December 1952 when a brown fog covered the city for about 5 days. This was precipitated by some very cold weather when every house stoked up its wood and coal fires. This, along with factory smoke, cars, and diesel fumes from the new double-decker buses, sprayed the city with ash and smoke. Also, London then had two coal-fired power stations, which added to the smog. The fog definitely turned brown. See

https://www.history.com/news/
the-killer-fog-that-blanketed-london-60-years-ago

In their stoic way, Londoners tried to ignore it and go about their business as best they could. However, the smog that had been created had a bad egg smell and eyes were stinging. Breathing wasn't too hot either. People's faces collected a black covering.

All transport except the underground railway, more or less came to a halt. Planes stopped. Cars were abandoned as drivers couldn't see where they were going. Buses survived for a while with conductors (the men who collected the money for bus trips and gave out tickets as receipts) leading the buses on their route with torches.

The smog was lethal. Cigarette smokers had trouble breathing. So did old and young people. Some 4,000 Londoners died during the smoggy period and more than twice that died in the following months.

Abigail

Brouncker died in 1684. He never married, but he was survived by his mistress Abigail Williams. His title passed on to his brother Henry who was considered to be the black sheep of the family. He was left virtually nothing in William's will though, "for reasons I think not fit to mention". Abigail inherited most of Brouncker's estate.

11.5 Blaise Pascal

Family

Blaise Pascal was born on the 19 June 1623 to Étienne and Antoinette Pascal. All that I know about his mother is that she died only three years after his birth. Blaise was the third child of three, the other two being Gilberte and Jacqueline. Étienne didn't marry again.

Even though Blaise strongly protested at the time, Jacqueline entered the Port-Royal Abbey in Paris. She died at 36. Gilberte was the elder of the two sisters and she married Florin Perier in 1641 and had at least four children. She sent her children to Port-Royal where they may have been taught by their aunt Jacqueline. Gilberte wrote biographies of both Pascal and Jacqueline and she assisted Blaise in two of his publications.

Étienne was a tax officer in Rouen when Blaise was born, but he sold his position in 1623 and moved the family to Paris. The profit from this transaction was invested in government bonds, which were meant to keep him prosperous for the rest of his life. One thing that linked Blaise's father and Fermat's was that Étienne was also a Noblesse de Robe.

The following quote may give you some idea of Blaise. By Adamson (1995), p. 15, Pascal was described as

> a man of slight build with a loud voice and somewhat overbearing manner. ... he lived most of his adult life in great pain. He had always been in delicate health, suffering even in his youth from migraine ... precocious, stubbornly persevering, a perfectionist, pugnacious to the point of bullying ruthlessness yet seeking to be meek and humble.

I think that you might overlook any possible sins there, as he did live in bad health and almost constant pain until his death. The migraines were really bad, but this wasn't the end of his suffering, as I'll tell you later.

(Blaise) Pascal's Theorem

All the Pascal children had above-average intelligence and Étienne decided to home-school them. It was clear early on that Blaise stood out. Among other things he wrote *Essay on Conics* when he was 16, which was related to work of Desargues on projective geometry. Descartes read it and couldn't believe that it had been written by Blaise. Étienne wasn't a slouch intellectually. There is a curve named after him (not Blaise), known as Pascal's limaçon. Consequently, Descartes thought that the *Essay on Conics* had been written by Étienne.

Pascal's result on conics is known as Pascal's theorem.

https://en.wikipedia.org/wiki/Pascal's_theorem.

To see what this looks like, draw six points A, B, C, D, E, F on a conic (such as the ellipse in Figure 11.3) and add the lines AE, FB which meet at P; CF, AD meeting at Q; and BD, EC meeting at R. Then $P, Q,$ and R are collinear (as the dashed line shows).

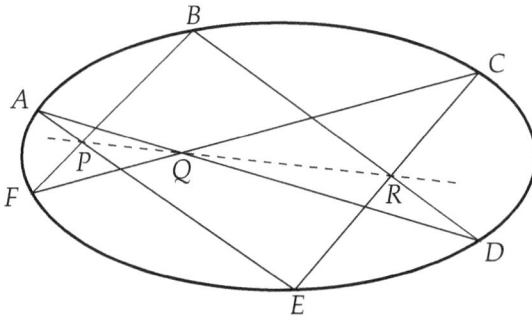

Fig. 11.3   Pascal's theorem

This theorem belonged to the newly created **projective geometry** of Desargues because the picture is "preserved by projection": projecting it onto another plane sends the conic section to another conic section, lines to lines, and intersections to intersections. Interestingly, the theorem also holds when the "conic" degenerates to a pair of straight lines. In fact, this special case of Pascal's theorem was discovered much earlier, by the Greek mathematician Pappus, whose proof of the isosceles triangle theorem we met in Section 2.3.

Puy-de-Dôme

Blaise actually did a lot of things that I don't have space here to tell you about. But it's worth saying that he did make one step forward in science. This followed on after Torricelli, who invented the barometer in 1643. He did this by taking a metre-long glass tube and filling it with mercury (see Figure 11.4). He then put his finger over the open end of the tube and inserted that end in a bowl of mercury. Perhaps surprisingly, the mercury didn't fill all of the inverted tube. It only stood about 76 cm above the mercury in the bowl. Left at the top of the tube was a vacuum. (Please don't make your own Torricelli barometer at home. Actually you may be more dissuaded by the cost of the mercury than my admonition.)

Unfortunately, Torricelli died in 1647, probably as the result of typhoid, but Pascal took up the idea. He constructed his own barometer and sent it to his brother-in-law Florin Perier who lived near the Puy-de-Dôme. This is a lava dome that is just under 1,500 m high. It's about 10 km from Clermont-Ferrand, where Pascal was born. The dome is certainly a popular place to visit because of the views of the city and the Puy range, a

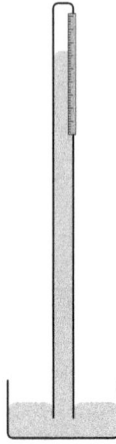

Fig. 11.4   Diagram of a barometer (from Wikipedia, Barometer)

sequence of mountains similar to Puy-de-Dôme. It's actually been the end of a stage of the Tour de France many times.

Florin Perier took several readings of his barometer at different places, from the bottom of the Puy-de-Dôme up to its top. The height of the mercury decreased the further up the mountain the measurements were taken. This had to be because the pressure/weight of the atmosphere decreased the higher up the mountain the measurements were taken. Hence, the air couldn't support as much weight/mass and the vacuum in the tube decreased.

If you have a barometer and a decent hill near you, you might like to repeat these experiments. Pascal himself did so on a less demanding climb in Paris. Today, the Eiffel Tower is a good place to experiment, if you happen to be passing through.

Pascal's Triangle

The triangle below was investigated by mathematicians in India and China long before Pascal was born, (see Wikipedia, Pascal's triangle and also Section 6.2), but Pascal discussed it systematically and was able to prove formulas and theorems about it. Perhaps that is good enough reason for the triangle of numbers to be named after him. You might wonder whether this was a bit hard on some of the people who had come before him though, so this section will also say something about Pascal's predecessors. Figure 11.5 shows the rows of the triangle from the zeroth to the seventh.

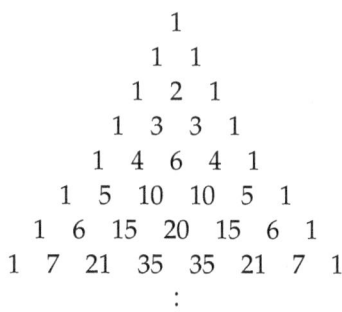

$$
\begin{array}{c}
1 \\
1 \quad 1 \\
1 \quad 2 \quad 1 \\
1 \quad 3 \quad 3 \quad 1 \\
1 \quad 4 \quad 6 \quad 4 \quad 1 \\
1 \quad 5 \quad 10 \quad 10 \quad 5 \quad 1 \\
1 \quad 6 \quad 15 \quad 20 \quad 15 \quad 6 \quad 1 \\
1 \quad 7 \quad 21 \quad 35 \quad 35 \quad 21 \quad 7 \quad 1 \\
\vdots
\end{array}
$$

Fig. 11.5 Pascal's triangle

As you can see, the leftmost and rightmost numbers in each row are 1 and every other number is the sum of the two numbers in the row above, to its left and right – a recurrence relation a little like that for the Fibonacci sequence. For example, the seventh row contains

$$1, \ 6 = 1 + 5, \ 15 = 5 + 10, \ 20 = 10 + 10, \ 15 = 10 + 5, \ 6 = 5 + 1, \ 1.$$

As a result, it's easy to compute each row from the one before, and the triangle goes on forever. Incidentally, you might find it interesting to see how many downward paths there are from the topmost 1 to the position of any given number, passing through numbers above it and moving from each position to either of the two positions below it.

Pascal's triangle is full of patterns (see Wikipedia, Pascal's triangle). For instance, the leftmost diagonal is all ones. The next diagonal is the whole numbers. The next diagonal is the triangular numbers. But what is the next diagonal that begins as 1, 4, 10, 20?

It's even possible to see the Fibonacci numbers if you are prepared to add the triangle numbers that are on parallel lines across the triangle. And why do the numbers in each row sum to a power of 2? To see the underlying reason for this rapid numerical growth in Pascal's triangle we need to review the idea of **permutations** and **combinations**.

Permutations, Combinations, and Induction

The theory of permutations and combinations arises from two questions, answered by Levi ben Gershon in 1321. See Katz *et al.* (2016), pp. 273–277.

(1) In how many ways can m different objects be arranged in order?
(2) In a set of m objects, how many subsets contain n objects?

The answer to Question 1 is $m(m-1)(m-2) \times \cdots \times 3 \times 2 \times 1$ because

the object in the first place may be chosen in m ways, after which

the object in the second place may be chosen in $m-1$ ways,

the object in the third place may be chosen in $m-2$ ways,

and so on. Thus, for each of the m choices for the first place there are $m-1$ choices for the second, making $m(m-1)$ choices for the first *two* places. Continuing this argument we find there are $m(m-1)(m-2)$ choices for the first *three* places, and so on, leading to

$m(m-1)(m-2) \times \cdots \times 3 \times 2 \times 1$ ways to order all m objects.

This is called the number of **permutations** of m objects, and we denote it by $m!$ for short, and pronounce the symbol "m factorial". (When I was a student, one of my teachers pronounced $m!$ as "m shriek"; but this never caught on.)

It is worth dwelling on the permutation argument a little longer, because it is a simple example of the important method of **mathematical induction**. I mentioned induction before, in Section 3.1, as another name for Euclid's "infinite descent", so I should explain why the two ideas are really the same. The method of "infinite descent" says that "infinite descent in the natural numbers is impossible"; namely, that any descending sequence of natural numbers is *finite*. Another way to say this is what you might call *finite ascent*; that any natural number may be reached from 0 by adding 1 a finite number of times.

This "finite ascent" principle is used in a style of induction proof involving two steps:

- a **base step**, which establishes that a certain statement holds for 0 (or some other starting value),
- an **induction step**, establishing that if the statement holds for any natural number k, then it holds for $k+1$.

These two steps together "complete the induction" because they imply that the property holds for any natural number m from the starting value up, since m may be reached by finitely often going from k to $k+1$.

Levi ben Gershon's argument really proves a statement about n:

in an ordering of m things,
the first n places can be ordered in $m(m-1)(m-2) \times \cdots \times (m-n+1)$ *ways*

and he proves it by induction on n. It is clearly true for $n = 1$, and if it is true for $n = k$ then it is true for $n = k + 1$, since that $(k + 1)$st place may be filled in $n - k$ ways. You may protest that the original argument is fine, if not clearer, but the induction step makes precise what I actually fudged when I said "and so on" in the penultimate line of the proof.

Most mathematicians would happily agree that it is good enough to say "and so on" here, but in other cases the induction step is not at all obvious, and it has to be spelled out. Levi ben Gershon was in fact somewhat aware of induction, though he lacked the notation to express it clearly. (You will notice that my proof relies on the ellipsis symbol, $\cdots$, which was not then in use. Euclid also suffered from the lack of such a symbol, and he fudged his induction arguments too.)

Anyway, having answered question 1, we can go on to question 2: in a set of m objects, how many subsets contain n objects? The number of such subsets was traditionally called the number of **combinations** of m things, "taken n at time". That phrase sounds quaint today, but we still use the term "combinations". To answer question 2 we first *order* the n things chosen from the m things (we will "disorder" them later). As before

the object in the first place may be chosen in m ways, after which

the object in the second place may be chosen in $m - 1$ ways,

the object in the third place may be chosen in $m - 2$ ways,

and so on, except that now we stop with the nth object, which can be chosen in $m - n + 1$ ways. Thus the number of ways to choose n objects from m, in order, is

$$m(m - 1)(m - 2) \times \cdots \times (m - n + 1), \quad \text{which equals} \quad \frac{m!}{(m - n)!}.$$

However, we don't care about the order of the n objects – all the $n!$ orderings of the n objects represent the same *set* – so we have to divide the number just obtained by $n!$, which gives

$$\frac{m!}{n!(m - n)!}$$

as the number of ways to choose n objects from a set of m objects. This number is denoted by $\binom{m}{n}$, and this symbol is pronounced "m choose n".

Our argument assumed that n was at least 1, in order to speak of the "first place". But of course there is also one set with zero members – the **empty set**. So we ought to have $\binom{m}{0} = 1$, which forces us to set $0! = 1$ if we want to keep the formula

$$\binom{m}{n} = \frac{m!}{n!(m - n)!}.$$

This is a weird property of 0, perhaps, but it turns out to be harmless. If 0 wants to have a factorial, 0! has to equal 1.

Combinations in Pascal's Triangle

The first theorem in Pascal's 1654 book on what he called the arithmetic triangle shows that the numbers in Pascal's triangle are none other than numbers of combinations. The theorem has a short proof by induction, but – as with many induction proofs – only if you are clever enough to guess what to assume in the induction step.

Pascal's Triangle Theorem: If we take the top row to be the zeroth row, and the leftmost position in each row to be the zeroth position, then the number in the mth row at the nth position from the left is $\binom{m}{n} = \frac{m!}{n!(m-n)!}$.

Proof: I am going to prove the theorem by induction on m, starting with $m = 0$. The formula is true for $m = 0$ because in that case the only value of n is also 0, and

$$\binom{0}{0} = \frac{0!}{0!0!} = 1,$$

thanks to our convention that $0! = 1$. This establishes the base step of the induction.

Next, for the induction step, I assume that the nth number in the mth row is $\binom{m}{n}$ for all $m \leq k$, and for all $n \leq m$. The aim is to prove that the same holds with $k + 1$ in place of k.

At this point I use the recurrence relation by which the numbers are constructed. The rule, remember, is that the nth number in row $k + 1$ is the sum of the $(n - 1)$st and nth numbers in row k. We are assuming that the latter numbers are $\binom{k}{n-1}$ and $\binom{k}{n}$, so we have to prove that

$$\binom{k+1}{n} = \binom{k}{n-1} + \binom{k}{n}.$$

This is true because

$$\binom{k}{n-1} + \binom{k}{n} = \frac{k!}{(n-1)!(k-n+1)!} + \frac{k!}{n!(k-n)!}$$

by the formula for combinations

$$= \frac{k!}{(n-1)!(k-n)!}\left(\frac{1}{k-n+1} + \frac{1}{n}\right)$$

taking a common factor from each term

$$= \frac{k!}{(n-1)!(k-n)!} \times \frac{k+1}{(k-n+1)n}$$

$$= \frac{(k+1)!}{n!(k-n+1)!}$$

$$= \binom{k+1}{n}.$$

This completes the induction step, so the result holds for all values of n, by mathematical induction. Q.E.D.

There is another way to see that the numbers in Pascal's triangle are numbers of combinations, without assuming the formula for combinations. In fact, this is how the triangle was discovered in China, some centuries before Pascal as we saw in Section 6.2, and it leads to an important result called the **binomial theorem**.

If you look carefully at Pascal's triangle you might be able to see some resemblance to the binomial expansions

$$(1+x)^2, \quad (1+x)^3, \quad (1+x)^4, \quad \text{and so on.}$$

Figure 11.6 shows them, from the zeroth power to the seventh.

$$
\begin{aligned}
(1+x)^0 &= 1 \\
(1+x)^1 &= 1+x \\
(1+x)^2 &= 1+2x+x^2 \\
(1+x)^3 &= 1+3x+3x^2+x^3 \\
(1+x)^4 &= 1+4x+6x^2+4x^3+x^4 \\
(1+x)^5 &= 1+5x+10x^2+10x^3+5x^4+x^5 \\
(1+x)^6 &= 1+6x+15x^2+20x^3+15x^4+6x^5+x^6 \\
(1+x)^7 &= 1+7x+21x^2+35x^3+35x^4+21x^5+7x^6+x^7
\end{aligned}
$$

Fig. 11.6 Powers of $1+x$

Yes, the coefficients of powers of x are none other than the numbers in Pascal's triangle! To be specific, the coefficient of x^n in $(1+x)^m$ is $\binom{m}{n}$. Why is this so? Well, after all, $(1+x)^m$ is the product of m factors of $1+x$:

$$(1+x)^m = (1+x) \times (1+x) \times (1+x) \times \cdots \times (1+x).$$

Each term in this product is the product of 1s from some factors, and xs from the others. Each time we choose x from n factors, and 1 from the others, we get a term x^n in the product. So the coefficient of x^n is the *number of ways* of choosing x from m of the factors. In other words, the coefficient is *the number of ways of choosing n things from m*; that is $\binom{m}{n}$. For this reason, the numbers $\binom{m}{n}$ are called **binomial coefficients**.

This connection between the numbers $\binom{m}{n}$ and $(1+x)^m$ also explains the recurrence relation

$$\binom{k+1}{n} = \binom{k}{n-1} + \binom{k}{n}.$$

The left side is the coefficient of x^n in $(1+x)^{k+1}$. And, since

$$(1+x)^{k+1} = (1+x)(1+x)^k = (1+x)^k + x(1+x)^k$$

we also have

coefficient of x^n in $(1+x)^{k+1}$

$=$ coefficient of x^n in $(1+x)^k$ + coefficient of x^{n-1} in $(1+x)^k$.

That is, $\binom{k+1}{n} = \binom{k}{n} + \binom{k}{n-1}.$

The Problem of Points

The binomial coefficients are the key to solving the problem of points that was introduced in Section 11.3. To keep things simple, let's suppose that two players, called H and T, agree to toss a coin 15 times. The first player to throw eight heads wins. Suppose also that the game is interrupted after eight tosses, at which stage five heads and three tails have occurred for H and T respectively. Thus player H will win if and only if at least three heads turn up in the remaining seven tosses.

The question is: in all possible continuations of the game, how many result in a win for H? There are $2^7 = 128$ possible continuations of the game, because there are two possibilities for each of the seven remaining tosses. H wins if heads occur in three or more times in these possible continuations. Now

- three heads can occur in $\binom{7}{3} = 35$ ways
- four heads can occur in $\binom{7}{4} = 35$ ways
- five heads can occur in $\binom{7}{5} = 21$ ways,
- six heads can occur in $\binom{7}{6} = 7$ ways,
- seven heads can occur in $\binom{7}{7} = 1$ way.

Thus H wins in $35 + 35 + 21 + 7 + 1 = 99$ out of 128 possible continuations of the game. This means that H should receive $99/128$ of the money the players have staked on the game.

Fermat solved the problem of points by adding numbers of combinations, as above. Pascal's triangle improves the solution by generating binomial coefficients quickly, though Pascal himself had a different solution.

The Pascaline

When Pascal was still in his teens his father, Étienne, had a financial contretemps with Cardinal Richelieu regarding bonds. This was not a good thing to have done and the father thought it best to discreetly leave Paris, but his children stayed. One day Jacqueline was taking part in a children's production. She performed quite well, apparently. Fortunately, Richelieu was at the play and enjoyed Jacqueline's performance so much that he gave Étienne a job as tax collector in Rouen.

Of course, there was a flip side to this. The Rouen office was in a mess! The major part of Étienne's job was cleaning up. This involved a large number of calculations and young Blaise was brought in to help. As a result, he made a calculator. This was called the Pascaline and it could do addition and subtraction. If they needed to do multiplication or division, they just had to repeat addition or subtraction. But it seems that Pascal had invented the first functioning machine. There had been another French calculator before this, but the machine got stuck if a string of carry overs was needed, like adding 9999 to something.

I'm sorry to say that the Pascaline, although it went into production, did not make a fortune for Pascal. Perhaps the world wasn't ready for it, or other people did not have such a need as Étienne did. However, there are still a few copies of the Pascaline in museums in Paris and Dresden.

Later Life

All of his life, at least from 18 onwards, Pascal suffered from bad health. He particularly suffered from extreme visual migraines. These produced regular headaches, including blindness on one side and hallucinations. These seem to have been responsible for two conversions to the Catholic Church. His first conversion was due to an accident. His horses bolted and swerved when he was driving his carriage. They left him hovering over the Seine. Although he wasn't hurt physically, he seems to have been affected psychologically and thought this was a sign from God.

A later religious experience included a fiery hallucination that Pascal thought had come from God.

As the result of these two incidents, Pascal virtually stopped doing mathematics and concentrated on thinking and writing about God. These notes were compiled in his work *Pensées* which are his thoughts on life and the Christian religion. They appear to be notes from which he was hoping to write an organised book, but he died before he was able to complete it. However, one of his notes was on a wager about God that I talk about in the next subsection.

Le Pari

This is about a wager (*le pari*), that Pascal wrote. At this time he was a religious man as the result of two conversions. Being a mathematician interested in probability and a Christian, it may not be surprising that Pascal wrote the wager. It can be found in his *Pensées* Part III, section 233. There has been much criticism of Pascal for this wager on the existence of God. But Pascal was a brilliant man and I think that he was partly in a leg-pulling mood. It's no coincidence that Pascal writes the wager as a discussion with de Méré. On the other hand, Pascal certainly wrote what Ore (1960), calls "a thought-provoking parable".

I've shortened *le pari* below from Ore (1960), without, I hope, destroying its substance.

> Pascal: God exists or he does not. Which side shall we take? Reason can decide nothing. An infinite chaos separates us. A game is being played where a decision, heads or tails, will be made at the end of this infinite distance. On what do you place your bet? By reason you cannot take one or the other; by reason you can defend neither choice. Therefore, do not blame the error of those who have made a choice, since you know nothing about it.

De Méré: No, but I blame them for having made a choice at all, not for their particular choice, for they are equally at fault, both he who chooses heads and he who chooses tails. The correct attitude is not to bet at all.

Pascal: Yes, but one is compelled to wager, it is not voluntary, you are in the game. Which side do you take? Let us see. Since you must wager, let us find out which alternative is the least profitable.

At this point, it is worth remembering that Pascal knows the concept of mathematical expectation and provides an argument of a wager for God.

Pascal: But here, actually, there is such an infinite life of infinite happiness to be won, one chance of winning against a finite number of possibilities for a loss, and that which you risk is finite. This eliminates all choice; whenever an infinite gain is involved and there is not an infinite number of losing chances against the winning ones, there is nothing to weigh, one must give all. And so, when forced to play, one must sacrifice reason to win life rather than to stake it against the infinite profit which may accrue just as easily as the loss: annihilation.

After more discussion in which de Méré is convinced by Pascal's argument, the conversation continues as follows.

De Méré: I confess, this I admit, but then is there no way of looking at the underside of the cards which have been dealt?

Pascal: Yes, the Holy Scriptures and the rest.

See more on this in

https://en.wikipedia.org/wiki/Pascal's_wager

How do you feel about Pascal's pari? Should it be criticised? Is it thought provoking?

Postscript on de Méré

In his early days at the court, de Méré appeared to be a really nice fellow, but his contact with Pascal certainly gave him airs. Part of a letter from de Méré to Pascal said:

Do you remember you once told me that you were no longer convinced of the excellence of mathematics? You write to me this time that I have disillusioned you completely and also that I have discovered things which you would never have perceived if you had not known me. I don't know, however, Monsieur, if you are as obliged to me as you may think. You still have the habit, which you have gathered from this science, not to judge anything except from your demonstrations, which are often false. These long reasonings drawn from line to line prevent you from obtaining the higher point of view which never deceives.

According to Ore (1960), Leibniz read the letter after it appeared in Bayle's *Dictionaire historique et critique* of 1697 (though Ore mispelled Bayle as "Boyle"). His response was

I almost laughed at the airs which the Chevalier de Méré takes on in his letter to Pascal.

It is a pity that De Méré believed he had made more than a passing contribution to mathematics. He might have been better to stick to his writing on philosophy, novels and poetry. Without him, mathematics would have eventually found someone who would have found and investigated the problems de Méré had sent to Pascal. Sooner or later, someone would have found what Pascal and Fermat found.

A Nasty Death

By an analysis of his letters and the biographies that his sisters wrote along with an autopsy (the first autopsy for forensic reasons was undertaken in 1302 in Bologna), some medical writers have suggested that Pascal suffered from coeliac disease. See

https://pubmed.ncbi.nlm.nih.gov/22258195/

Coeliac, an immune reaction to gluten, would have been an incredibly bad thing to get in the mid-17th century, when no one knew how to cure or relieve it. It is likely that the autopsy showed damage to the lining of Pascal's intestine. This is likely to be painful, possibly very painful, and lead to side effects such as anaemia, bloating and diarrhoea. His last years of life could have been spent better in another body.

It is also suggested that his death was in part caused by a subdural haematoma. Whatever the cause, Pascal died a very painful death on 19 August 1662 aged 39, after some three years of exceptionally poor health.

I must say that Pascal's death has made me really quite sad. You don't expect mathematical heroes to die so badly.

11.6 René Descartes

The Name

Everyone who has done high school mathematics will know the name "Descartes", though perhaps without connecting "Descartes" with "Cartesian". The latter name is the adjective for the coordinate system that shows geometry in an algebraic setting. I plan to go into this in greater detail in the next chapter. But I should add here that Fermat may have well lost the chance to have been the adjective for coordinate systems and not Descartes. Fermat certainly seemed to beat Descartes time-wise, but he filed his work away and it was only found after his death.

Oh, there's one more thing I want to say here. Descartes has a well-known saying: *"Cogito, ergo sum"*. For non-Latin speakers, "I think, therefore I am". This was the starting point of Descartes' work in philosophy, for which he is also famous.

Early Life

Descartes was born in France at a place called La Haye which is not that far off the main highway between Tours and Poitiers. However, you may not be able to find this town on your SAT NAV when you are exploring the French countryside. This is because it is now called "Descartes". However, there is a new La Haye. It was established in 2016 as a merger of several small places. It is a long way from Descartes!

The birthday of this peripatetic mathematician and philosopher was 31 March 1596 and he was baptised as a Roman Catholic a few days later. Again, like pretty well every mathematician in this book, he was born into a wealthy family. His father graduated in law and held a position of counsellor at the Parlement of Brittany. But did Descartes become independently wealthy himself? He certainly managed to find the money for a great deal of travelling, so he must have had access to a reasonable source of cash. Well, it turns out that in 1623 he went back to La Haye and sold all of his property so that he could invest in bonds. Unlike Étienne Pascal, Descartes wasn't later harmed financially by his investment. So, like most of the mathematicians in this book, Descartes lived a comfortable life.

Between 1607 (though the year is not certain), and 1614, Descartes attended the Jesuit La Flèche College in Anjou. There he managed to study maths, which in those days covered arithmetic, geometry, music, and astronomy. As a boarder, Descartes and all the other children, were forced to get up at 5 in the morning! But he managed to avoid this by being a sickly child. Consequently, he was allowed to stay in bed until 11 am. This seemed to be a good way to live and so he continued the practice for the rest of his life. Though I'm not sure how he got away with this while he was a soldier. However, it is possible that the enemy thought that they couldn't wake up poor Descartes so they didn't attack before 11:15!

Actually, when he wasn't soldiering, he usually spent the mornings working in bed. On rising he would spend time relaxing until the evening when he engaged in writing letters. These tended to be about the ideas he was considering at the time.

By 1616 he had graduated in civil and canon law at the University of Poitiers. This study wasn't what he really wanted to do; rather it was what his father wanted him to do. After completing his university education, he went to Paris to begin his travels around Europe.

Fig. 11.7   René Descartes (from Wikipedia, Descartes)

Description of Descartes

What did he look like? Well, you can get some idea of this by looking at the portrait in Figure 11.7 (based on one made from life by the Dutch artist Frans Hals). But what was he like as a person? The next two quotes from Langer (1937), p. 509, and from Russell (1961), p. 544, may give you some idea.

> In appearance Descartes was a small man of rather slight figure with a large head. His nose was prominent, his lower lip somewhat protruding, his beard and moustache of a semi-military type, and his hair growing down upon his forehead almost to his eyebrows. He wore a wig of natural colour to which he always gave fastidious attention, as he did also to his clothes which were now invariably of black cloth. In demeanour he was generally cheerful, rarely gay. His manners were always refined, gentle, and polite, and his temper tranquil and easy. As a personality he was proud, somewhat aristocratically reserved, sensitive, a bit angular, and, though a shade domineering, was pre-eminently obliging.

That doesn't sound too bad.

> He always was well dressed, and wore a sword. He was not industrious; he worked short hours, and read little. When he went to Holland he took few books with him, but among them were the Bible and Thomas Aquinas. His work seems to have been done with great concentration during short periods; but perhaps, to keep up the appearance of a gentlemanly amateur, he may have pretended to work less than in fact he did, for otherwise his achievements seem scarcely credible.

But I'm not sure about the sword. Was this for defence or to impress people? Given "his temper tranquil", the sword was maybe more for effect.

Now Descartes may have felt himself to be a self-made philosopher because he worked on the subject from scratch, starting with *"Cogito ergo sum"*. This approach did get him in conflict with the more classic side of the subject – but what philosopher *doesn't* conflict with his predecessors?

Descartes' Travels

Over the next 30 years or so, Descartes travelled around Europe.

1618.   Descartes joined the Dutch States Army at Breda, southern Netherlands, as a mercenary. This army was led by Maurice of Nassau. On the death of Maurice's brother, Maurice became Prince of Orange, and had control over almost all of the Dutch Republic. This was during the Twelve Years of Truce in the 80 years' war with Spain.

1619. He then served under Maximilian of Bavaria, presumably because there was no fighting going on under Maurice. The following year he was present at the Battle of the White Mountain near Prague.

During November of this year, to keep warm, Descartes slept in a room with a fire. While there, he had three dreams that were to set a path for his future investigations. After waking, he supposedly laid the foundation of his work on analytic geometry and had the idea of approaching philosophy via mathematical methods. This may well have been the beginning of analytic geometry, but we have no paper trail to confirm it.

At this point, he seems to have turned his back on wars. It is likely that he would have then returned to Paris, but there was a plague there so he went on to a number of other places.

1620. He went to Bohemia.

1621. Hungary.

1622–1623. Germany, Holland and France.

1623. He arrived back in France and contacted Mersenne, whom we met in Section 3.6. He also met Claude Mydorge who was an expert on conic sections. This might have led Descartes towards showing that conics are the curves given by quadratic polynomials. Then Descartes spent some time in Switzerland and Italy (Venice and Rome).

1625. Descartes returned once again to Paris where his home became a meeting place for philosophers and mathematicians.

1627. He was at the siege of la Rochelle under the leadership of Cardinal Richelieu. The conflict there was between the Huguenots, protestants of la Rochelle, and the forces of Louis XIII of France. I can't resist including the stunning painting of the siege, by Henri Motte in 1881, showing Richelieu in red in the foreground (Figure 11.8).

1628. Descartes apparently found Paris to be a bit claustrophobic and he looked for a quiet place in Holland, where many thinkers sought refuge from intolerant regimes. Perhaps Amsterdam was to his liking. However, he did move round Holland a reasonable amount, even going beyond, to Paris, a couple of times.

In 1629, he went to Franeker (120 km along the coast to the North of Amsterdam), where he attended the University of Franeker. There he studied maths and astronomy.

Later he went to Amsterdam, where he had an affair with a servant. As a result, he had a daughter, Francine. He was clearly fond of her because he is said to have felt the greatest sorrow of his life when she died at the age of 5.

Fig. 11.8   The siege of la Rochelle (from Wikipedia, Siege of La Rochelle)

In the twenty years or so that Descartes lived in Holland, he produced most of his publications. These were:

Rules for the Direction of the Mind – written in 1628 but not published until 1701.

Optics

Meteorology

Treatise on Man

Geometry, Descartes (1637) – this book showed his connections between geometry and algebra and hence was a basis for analytic geometry.

He spent four years on his *Treatise of the World* and in 1633 he was about to have it published. However, Galileo had already published his *Dialogue on the Two World Systems*. This was a comparison of Claudius Ptolemy's and Copernicus' work. The Inquisition of Italy then summoned Galileo to Rome where he was found guilty of heresy in a rigged trial. This made Descartes decide to hold off his publication, concerned he would suffer a similar fate. He never published the treatise, but it is believed that *Optics* and *Meteorology* were originally meant to be part of it.

1649. He was invited to Stockholm to work with Queen Christina. She was an early riser and demanded to work at five in the morning. This was against his lifetime work habits, but he went on demand. In the winter there was snow and the weather was cold. It seems that he caught pneumonia and died. There were rumours that he may have been poisoned.

The queen had begun to tire of their meetings, but there is no evidence that she might have been involved in such a dramatic and evil disposal.

Chapter 12

Numbers and Curves

The ancient Greeks studied many curves apart from the circles that are in Euclid's *Elements*. The most important of these are the **conic sections** – the ellipses, parabolas, and hyperbolas obtained by cutting a cone with a plane (Figure 12.1).

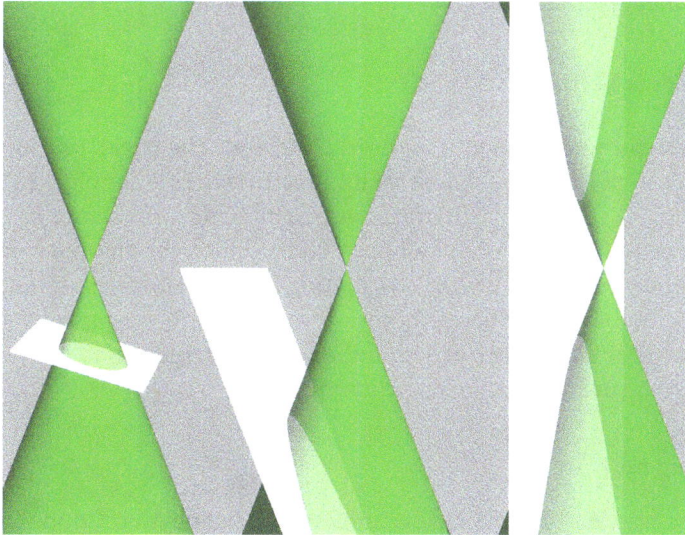

Fig. 12.1   The conic sections

The conic sections were known before Euclid. 50 years after Euclid, Apollonius made the most thorough investigation of them, producing hundreds of theorems about these curves. It was by studying Apollonius that Fermat (and, apparently independently, Descartes) made one of the biggest breakthroughs in the history of mathematics – the use of **coordinates** and **equations** to study curves.

What was it about Greek mathematics that prompted this discovery? And why was the time ripe for it to occur around 1630? Historians argue about these things, but I am fairly confident that the respective answers are:

- Greek geometry of conic sections was ripe for an algebraic treatment, because it related these curves to rectangles, and rectangles correspond to *products* (of lengths).
- Around 1600, algebra became as efficient as arithmetic with the use of symbols for unknowns, arithmetic operations, and equality.

In this chapter I'll first sketch the main ideas about curves in ancient Greek mathematics, especially those that were ripe for translation into algebra, and also the algebraic advances of the 16th century, which made algebra ready for use in geometry. Then I'll describe some geometric advances that algebra made possible.

12.1 Conic Sections

The conic sections shown in Figure 12.1 include the circle, of course, but the circle is special in the sense that it can be drawn by a simple instrument, the compass. The Greeks did not have instruments for drawing parabolas, hyperbolas, and non-circular ellipses, so they thought the next best thing was to imagine them "drawn" by cutting a cone. However, though they fell back on the cone to bring these curves into being, they mostly studied the behaviour of each curve within the *plane* that contains it. This was the first step towards the use of coordinates.

Doubling the Cube

Another way in which conic sections were "the next best thing" for the Greeks: they used them to solve problems they could not solve with ruler and compass. A spectacular example is the problem of doubling the cube, which was solved by Menaechmus (fourth century BCE) by intersecting a parabola with a hyperbola. Menaechmus is credited with the discovery of conic sections, so he might even have introduced them for this purpose.

The solution is easy with modern equations for the parabola and the rectangular hyperbola. Remember from Section 2.5 that the problem of doubling the cube amounts to finding a length x such that the volume x^3 of the cube of side x is twice the volume of a unit cube, so $x^3 = 2$. Menaechmus solved this problem by finding the intersection of the parabola we would describe as $y = x^2$ with the hyperbola $y = 2/x$ (Figure 12.2).

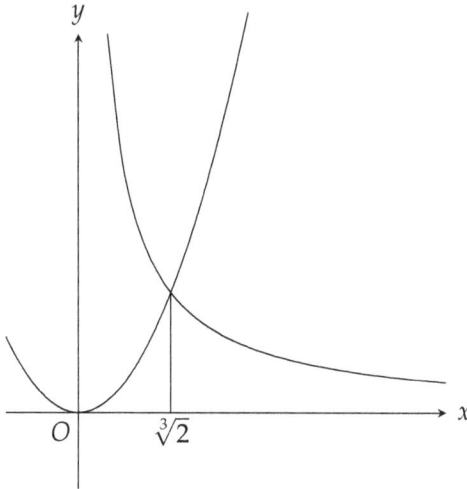

Fig. 12.2 Finding $\sqrt[3]{2}$

Algebraic symbolism makes it easy to see why this works. Intersection occurs where

$$y = x^2 = 2/x, \quad \text{that is, where} \quad x^3 = 2, \quad \text{so} \quad x = \sqrt[3]{2}.$$

But how might Menaechmus have been able to simulate this algebra in the Greek way, by comparing areas? It is easy to describe the area property of the hyperbola: $y = 2/x$ is the same as $xy = 2$, which simply says the rectangle between the hyperbola and its asymptotes has the constant area 2 (Figure 12.3). The hyperbola is said to be the **locus** (from the Latin word for "place") of points P with this property.

This assumes that Menaechmus was aware of the asymptotes to the hyperbola (our x- and y-axes here), which seems reasonable. Likewise, it seems reasonable to assume the axis of symmetry of the parabola to be known, and hence also its perpendicular at the base of the parabola. Then our equation $y = x^2$ of the parabola is equivalent to the equality of the square and the rectangle shown in Figure 12.4. So the parabola is the locus of points P for which these areas are equal.

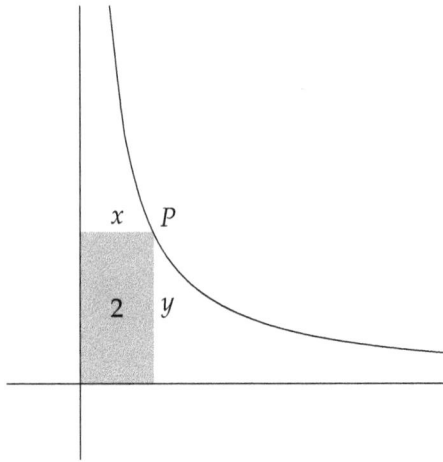

Fig. 12.3 The area property of the rectangular hyperbola $y = 2/x$

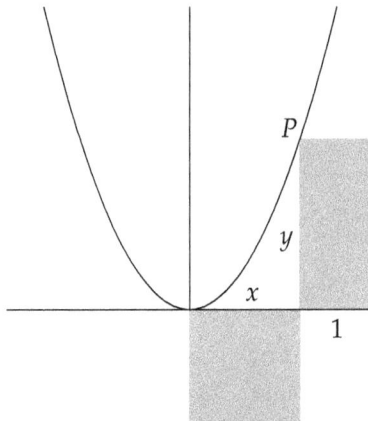

Fig. 12.4 The area property of the parabola $y = x^2$

It is so easy to define this parabola and hyperbola in terms of squares and rectangles that you may wonder why it was ever deemed necessary to refer to sections of the cone – except of course as an interesting alternative way to describe these curves. The best answer, in my opinion, was given by Heath (1896) in his English edition of the *Conics* of Apollonius, p. xcvii:

> Greek geometers in general seem to have connected the conic sections with the cone only because it was in their view necessary to give the curves

a geometric definition expressive of their relation to other known geometrical figures, as distinct from an abstract definition as the loci of points sat-
isfying certain conditions . . . Menaechmus and his contemporaries would
perhaps hardly have ventured, without such a geometrical definition, to
regard the loci . . . as being really curves. When however they were found
to be producible by cutting a cone in a particular manner, this fact was a
sort of guarantee that they were genuine curves . . .

In other words, the Greeks preferred a *construction* to a *defining property*
(as was true of the equilateral triangle in Proposition 1 in Euclid's *Elements*). As long as there was no efficient means of working with defining
properties – that is, the algebra of equations – this was understandable.
But to explain conic sections to a modern audience, equations win hands
down. They already did in the 17th century.

The Conics *of Apollonius*

The ancient master of conic sections was Apollonius of Perga, who lived
from approximately 240 BCE to 190 BCE. Following up on the work of
Menaechmus, Euclid, and Archimedes, Apollonius proved almost everything one would wish to know (if not more) about conic sections in the
387 propositions of his *Conics*. From the 17th century onwards, European
mathematicians learned about conic sections from editions of Apollonius.
In fact, it was after studying Apollonius in the late 1620s that Fermat hit
on the idea of describing curves by means of x and y coordinates. He and
Descartes independently noticed that the conic sections were the curves
given by equations of second degree in x and y.

This simultaneous discovery is not completely surprising. The language of squares and rectangles used by Apollonius translates into equations of second degree as soon as one thinks to describe the lengths of
sides by x, y, and constants. The hard part is to show that *any* equation of
second degree, namely

$$ax^2 + bxy + cy^2 + ex + fy + g = 0,$$

can be converted to one of the standard forms, such as $y = x^2$ for the
parabola. This is hard to do by slicing and dicing squares and rectangles –
it is only feasible in a good algebraic symbolism. We will see how algebra
reached this state just in time for Fermat and Descartes in the next section.

The other thing to bear in mind is that even mathematicians like to stick
with what they know. Until the 17th century, there was no good reason to

object to the Greek way of doing things – and the Greeks could do geometry more easily than algebra. The view of Heath about "genuine curves", quoted above, is borne out in Book I, Proposition 52 of Apollonius' *Conics*. This proposition is about "finding" a parabola, given its definition by equality of a rectangle and a square, like that shown in Figure 12.4. Today, we feel that $y = x^2$ tells us all we need to know about the parabola, but this view took hold only in the 17th century. What Apollonius "finds" is in fact a cone and the plane that cuts it in the given curve. His Propositions 54 and 56 similarly "find" cones and planes for a hyperbola and ellipse that are similarly "given" by equations between squares and rectangles.

12.2   Algebra

Algebra, as we saw in Section 8.4, got its name from al-Khwarizmi's book around 800 CE, along with the intuition that algebra was about "restoring and balancing". In fact, for centuries after that, "algebra" also meant the art of resetting broken bones, and some dictionaries still include bone-setting as one meaning of the word. We also saw that al-Kwarizmi interpreted the objects being equated and manipulated as geometric things such as squares and rectangles. This idea, too, persisted until the 16th century.

In all this time, alas, algebra made very little progress. The ancient Babylonians could solve quadratic equations, and so could everyone else from Euclid to al-Khwarizmi. But there was no systematic way to solve equations of higher degree. Menaechmus had solved the equation $x^3 = 2$, in a sense, by intersecting a parabola and a hyperbola, but there was no *formula* for solving cubic equations, comparable to the formula for solving quadratics that had been stated centuries earlier by Brahmagupta (see Section 8.3).

Algebra finally broke out of the quadratic prison in the early 16th century, when the Italian mathematician Scipione del Ferro discovered a formula (involving cube roots and square roots) for solving a large class of cubic equations. The formula was kept secret for some time, but rediscovered in the 1530s by Niccolò Fontana (better known by his nickname, Tartaglia) and eventually divulged to Girolamo Cardano, who published it in a book known as the *Ars magna* (Great art) in 1545. The *Ars magna* gave formulas solving all types of cubic equations, as well as a solution of *fourth* degree equations discovered by Cardano's student Lodovico Ferrari.

Figure 12.5 shows the title page of the *Ars magna*, including a portrait of Cardano. You will notice that the title is *not* "Ars magna" but the longer expression *Artis magnae sive de regulis algebraicis* (meaning something like "On the great art, or algebraic rules"), however "Ars magna" is the short form by which it is generally known.

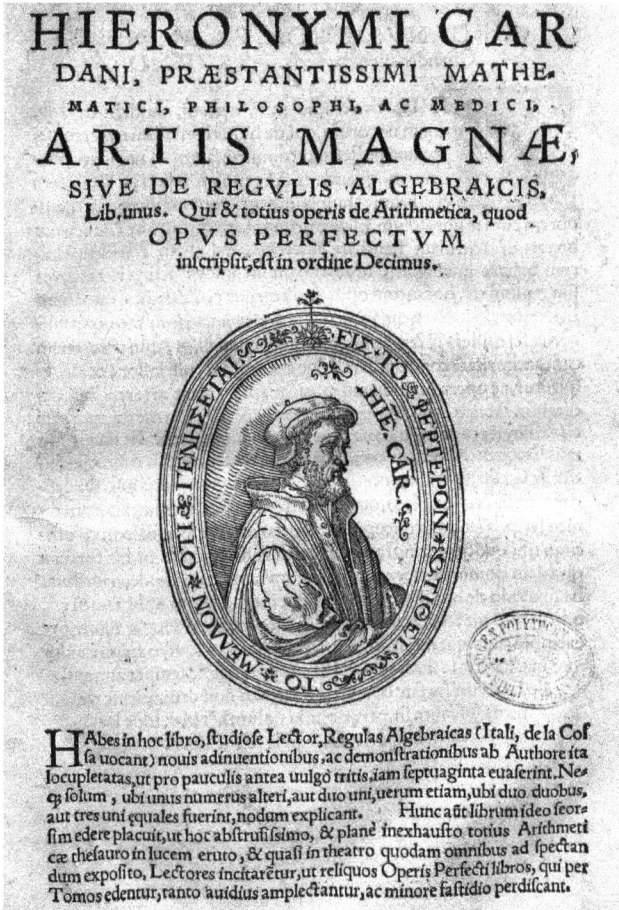

HIERONYMI CAR
DANI, PRÆSTANTISSIMI MATHE-
MATICI, PHILOSOPHI, AC MEDICI,
ARTIS MAGNÆ,
SIVE DE REGVLIS ALGEBRAICIS,
Lib.unus. Qui & totius operis de Arithmetica, quod
OPVS PERFECTVM
inscripsit,est in ordine Decimus.

Habes in hoc libro,studiose Lector,Regulas Algebraicas (Itali, de la Cos
fa uocant) nouis adinuentionibus,ac demonstrationibus ab Authore ita
locupletatas,ut pro pauculis antea uulgò tritis,iam septuaginta euaserint. Ne-
q; solum , ubi unus numerus alteri,aut duo uni,uerum etiam,ubi duo duobus,
aut tres uni æquales fuerint,nodum explicant.   Hunc aût librum ideo seor-
sim edere placuit,ut hoc abstrusisimo, & plane inexhausto totius Arithmeti
cæ thesauro in lucem eruto,& quasi in theatro quodam omnibus ad spectan
dum exposito, Lectores incitarêtur,ut reliquos Operis Perfecti libros, qui per
Tomos edentur,tanto auidius amplectantur,ac minore fastidio perdiscant.

Fig. 12.5   Cardano and the *Ars magna* (from Wikipedia, *Ars Magna* (Cardano book))

Cubic Equations

The gist of the solution of cubic equations can be illustrated in the case of the following equation (which is not actually a special case, since all cubics may be reduced to this form):

$$x^3 = px + q. \qquad\qquad (*)$$

The trick is to set $x = u + v$, in which case

$$\begin{aligned}
x^3 &= (u + v)^3 \\
&= u^3 + 3u^2v + 3uv^2 + v^3 \\
&= 3uv(u + v) + (u^3 + v^3) \\
&= 3uvx + (u^3 + v^3).
\end{aligned}$$

This expression can be made equal to $px + q$, as required by (*), if

$$3uv = p \quad \text{and} \quad q = u^3 + v^3 \qquad\qquad (**)$$

You might think that solving (**) is no simpler, but let's carry on. The first equation in (**) gives $v = p/3u$, and substituting this for v in the second equation of (**) gives

$$u^3 + \left(\frac{p}{3u}\right)^3 = q, \quad \text{or} \quad u^6 + \left(\frac{p}{3}\right)^3 = qu^3,$$

which is a *quadratic* equation for u^3, namely

$$(u^3)^2 - qu^3 + \left(\frac{p}{3}\right)^3 = 0.$$

Aha! The solutions of this equation for u^3, by the quadratic formula, are

$$u^3 = \frac{q + \sqrt{q^2 - 4\left(\frac{p}{3}\right)^3}}{2} \quad \text{and} \quad u^3 = \frac{q - \sqrt{q^2 - 4\left(\frac{p}{3}\right)^3}}{2},$$

which have the slightly tidier form

$$u^3 = \frac{q}{2} + \sqrt{\left(\frac{q}{2}\right)^2 - \left(\frac{p}{3}\right)^3} \quad \text{and} \quad u^3 = \frac{q}{2} - \sqrt{\left(\frac{q}{2}\right)^2 - \left(\frac{p}{3}\right)^3}$$

If we now go back to (**) and solve instead for v^3, we will get the same two possible values for v^3. And since $u^3 + v^3 = q$ from (**), if we take either solution to be u^3, then the other must be v^3. Taking cube roots, to get u and v, finally gives the solution of (*):

$$x = u + v = \sqrt[3]{\frac{q}{2} + \sqrt{\left(\frac{q}{2}\right)^2 - \left(\frac{p}{3}\right)^3}} + \sqrt[3]{\frac{q}{2} - \sqrt{\left(\frac{q}{2}\right)^2 - \left(\frac{p}{3}\right)^3}}. \qquad (***)$$

Algebraic Symbolism

Even when the solution of the cubic is written in modern symbolism, you've got to be impressed by its cleverness. It is even more impressive when you realise that the Italians did it the hard way, using geometric manipulation in the style of al-Khwarizmi. Cardano, too, thought it was pretty impressive, as he told his readers:

> In our own days Scipione del Ferro of Bologna has solved the case of the cube and first power equal to a constant, a very elegant and admirable accomplishment. Since this art surpasses all human subtlety and the perspicuity of men's minds, whoever applies himself to it will believe that there is nothing that he cannot understand.
>
> Cardano (1545), p. 8.

The solution of the cubic was not only a big morale booster for algebra but also a stimulus for further progress. In the 1550s Cardano's compatriot Rafael Bombelli began work on a new algebra book, *L'Algebra*, which made improvements in several directions. One was the introduction of **complex numbers**, which he used to brilliantly reconcile two apparently different solutions of the equation $x^3 = 15x + 4$.

You see, the Cardano formula (***) with $p = 15, q = 4$ gives

$$x = \sqrt[3]{2 + 11\sqrt{-1}} + \sqrt[3]{2 - 11\sqrt{-1}}$$

or

$$x = \sqrt[3]{2 + 11i} + \sqrt[3]{2 - 11i},$$

as we would write it today, with $i = \sqrt{-1}$. On the other hand, you can see that $x = 4$ satisfies $x^3 = 15x + 4$. How can these two things be equal? Bombelli made the inspired guess that $(2 + i)^3 = 2 + 11i$ and $(2 - i)^3 = 2 - 11i$, which can be verified if we assume i behaves like an ordinary number and $i^2 = -1$.

With this assumption, *expanding the number concept to include* $\sqrt{-1}$, we find that the Cardano formula solution indeed equals 4, because

$$x = \sqrt[3]{2 + 11i} + \sqrt[3]{2 - 11i}$$
$$= \sqrt[3]{(2 + i)^3} + \sqrt[3]{(2 - i)^3}$$
$$= (2 + i) + (2 - i) = 4.$$

Bombelli's decision to treat $\sqrt{-1}$ as a genuine number had reverberations far into the future, but *L'Algebra* also made progress towards simplifying algebraic calculation. For example, he wrote exponents as superscripts, and used single symbols for "plus" and "minus". In particular he wrote the equation $x^3 = 15x + 4$ as

$$1\underline{3}. \text{ eguale a } 15\underline{1}.\text{p}.4.$$

This symbolism rather cleverly manages without a symbol x for the unknown, but it won't work too well if there are two or more unknowns.

Letters for unknowns in equations were introduced by the French mathematician Francois Viète around 1600 and independently by Thomas Harriot in England. Harriot also introduced the signs $<$ and $>$ we use today and our signs for square and cube roots. Strangely, the $=$ sign, introduced by the Englishman Robert Recorde in 1557, did not catch on until well into the next century. This was despite Recorde's sensible justification for the sign consisting of two lines – "bicause noe .2. thynges can be moare equalle" – and the symmetry of the sign, which competing signs lacked. Another oddity of early algebraic notation was that x^2 was written xx, even though the higher powers were written $x^3, x^4, x^5, \ldots$.

Nevertheless, by the time Fermat was studying Viète in the late 1620s, there was an algebraic notation that made algebraic *computation* as efficient as computation with numbers. This made algebra ripe for use in geometry. In fact, algebra and geometry were ready to change places: algebra in al-Khwarizmi's time depended on geometry, but from the 1630s onward, geometry would depend on algebra.

12.3   Algebraic Geometry

Once it was realised that equations were a good way to represent curves it was easy to rewrite the Greek descriptions of conic sections as quadratic equations. After that, the next question was: does every quadratic equation in x and y represent a conic section? This was a harder problem, which could not have been contemplated until efficient algebraic machinery was available. Fermat and Descartes apparently first solved it, independently, around 1630. Let's see what was involved.

The general quadratic equation is

$$ax^2 + bxy + cy^2 + dx + ey + f = 0, \quad \text{where } a, b, c, d, e, f \text{ are numbers.}$$

This equation refers the curve to two lines, the x- and y-axes, and the aim is to find new axes for which the curve's equation has a simpler form. The forms we are looking for are

$$y = x^2 \qquad \text{(parabola)}$$

$$px^2 + qy^2 = r \quad \text{where } p, q, r > 0 \qquad \text{(ellipse)}$$

$$px^2 - qy^2 = r \quad \text{where } p, q > 0 \qquad \text{(hyperbola)}$$

Viewing a curve relative to several different lines is actually an old idea: Apollonius did it a lot, but algebra gives a better way to track the changing view.

Here is an example: the hyperbola $xy = 1$. We know that the x- and y-axes in this case are the *asymptotes* of the hyperbola, so the equation reflects the view of the curve from its asymptotes. Now suppose we view the curve relative to its two axes of symmetry, which we might call the X- and Y-axes (Figure 12.6).

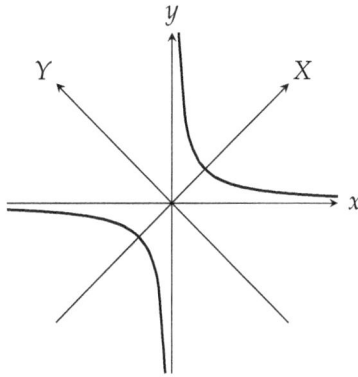

Fig. 12.6   Hyperbola and two sets of axes

The change of axes is reflected by equations relating the old coordinates to the new, which happen to be

$$x = \frac{X}{\sqrt{2}} - \frac{Y}{\sqrt{2}},$$

$$y = \frac{X}{\sqrt{2}} + \frac{Y}{\sqrt{2}}.$$

With this change of coordinates the old equation $xy = 1$ becomes

$$\left(\frac{X}{\sqrt{2}} - \frac{Y}{\sqrt{2}} \right) \left(\frac{X}{\sqrt{2}} + \frac{Y}{\sqrt{2}} \right) = 1, \quad \text{that is,} \quad \frac{X^2}{2} - \frac{Y^2}{2} = 1,$$

which is in standard form for a hyperbola.

In general, the xy term in the general quadratic equation may be removed by *rotation of axes* through some angle (not necessarily the $45°$ rotation used here). The x and y terms may be removed by *shifting the origin*, which means changing x and y to $x + p$ and $y + q$ respectively, for some numbers p and q. These coordinate changes suffice to produce an equation in one of the standard forms, apart from some extreme cases, which may be considered "degenerate" conic sections.

For example, removing an xy term could give the equation $X^2 - Y^2 = 0$, which is *not* a hyperbola. Instead, it is $(X - Y)(X + Y) = 0$, which represents the pair of straight lines $X = Y$ and $X = -Y$. However, if you think about it, any pair of intersecting lines lie on a cone. Even worse, we could get a pair of parallel lines. But in this case they lie on a cylinder – which can be considered as a "degenerate" cone.

Revisiting the Greek Construction Problems

The algebraic view of straight lines and circles gave new insight into the ancient idea of ruler and compass constructions, as Descartes seems to have been the first to realise. Instead of drawing a line between given points $A = (a_1, a_2)$ and $B = (b_1, b_2)$ one can write down the equation

$$\frac{y - a_2}{x - a_1} = \frac{b_2 - a_2}{b_1 - a_1},$$

which says that the slope between any point (x, y) on the line and the particular point A is the same as the slope between A and B. This (linear) equation in x and y is satisfied by precisely the points (x, y) on the line, so we call it the equation *of* the line.

Similarly, instead of drawing the circle with centre $C = (c_1, c_2)$ and radius r, one can write down the equation of the circle,

$$(x - c_1)^2 + (y - c_2)^2 = r^2,$$

which says that the distance from C to any point (x, y) on the circle is the constant r. This (quadratic) equation in x and y is the equation *of* the circle.

A ruler and compass construction proceeds from given points, drawing lines and circles to obtain further points where these lines and circles intersect, then using the new points to draw further lines and circles, and so on. If instead we form the equations of the initial lines and circles, the further points are found by finding common solutions of pairs of linear and quadratic equations. For example, to find the intersection of two lines, we find the solution of two linear equations. We can do this using the *rational* operations of addition, subtraction, multiplication, and division.

Intersecting the line $y = mx + c$ and the circle $(x - c_1)^2 + (y - c_2)^2 = r^2$ amounts to solving the quadratic equation

$$(x - c_1)^2 + (mx + c - c_2)^2 = r^2$$

for x, and hence may also involve square roots, as well as the rational operations. The same is true for the exceptional lines $x = d$. Finally, the intersection of two circles may also be found by solving a quadratic equation. This is not so obvious, but notice that if you write down the equations of the two circles and expand them, then subtracting one from the other will eliminate the x^2 and y^2 terms.

To sum up, the coordinates of all the points arising in a ruler and compass construction may be found from the given points by rational operations and square roots. This means we can expect to construct numbers like $\sqrt{3}$ or $\sqrt{1 + \sqrt{5}}$ or $\sqrt{\sqrt{3} + \sqrt{5}}$, or even more complicated numbers, but always with nested *square* roots. What about the number $\sqrt[3]{2}$?

As we saw in Section 12.1 this is the number needed to solve the problem of **doubling the cube**. So, doubling the cube *by ruler and compass* is possible only if $\sqrt[3]{2}$ is expressible by nested square roots. This situation may remind you of the one faced by Fibonacci, and described at the end of Section 9.6. Remember, he had a number that satisfied a cubic equation, and he claimed that this number could not be expressed by *certain* nested square roots, which satisfy equations of degree 2, 4, or 8.

Well once again we have a number $\sqrt[3]{2}$ that satisfies a cubic equation, and now we are asking whether it can be expressed by any number of nested square roots. Such an expression will satisfy an equation of degree 2, 4, 8, or any power of 2 – but seemingly *not* an equation of degree 3. This is true but, as we mentioned in the case of Fibonacci's number, it was not proved until the theory of fields was developed in the 19th century. This is how the problem of doubling the cube was finally settled. The problem of **trisecting the angle** was settled (negatively) in the same way, because trisecting the angle also amounts to solving a cubic equation.

Curves of Higher Degree

The conic sections make a good introduction to algebraic geometry, but we probably don't learn anything about them that Apollonius didn't already know, even if we can prove theorems more easily. The algebraic method doesn't really shine until we apply it to curves of higher degree, which the Greeks could barely touch.

The Greeks looked at a handful of curves that we now know to be of degree 3 and 4, but without algebra they could not study them systematically. For the history of some of these curves, see Brieskorn and Knörrer (1981), Chapter 1. Once algebra is available, the exploration of **cubic** (degree 3) curves reveals a host of interesting new features. Some of them have names that have become part of everyday speech, namely, **inflection points** and **cusps**.

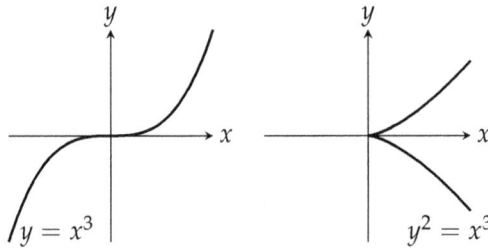

Fig. 12.7   Inflection point and cusp

These occur in the curves $y = x^3$ and $y^2 = x^3$ (see Figure 12.7). An inflection point, where the bend of the curve changes direction, probably corresponds quite well to the everyday meaning. A cusp, which is a sharp point like the one at the origin of $y^2 = x^3$ is, I suspect, not exactly what people usually mean when they say they are "on the cusp of ...".

Other interesting features that occur with cubic curves are **self-crossings** and **separate components**. These occur with the curves $y^2 = x^2(x + 1)$ and $y^2 = x(x^2 - 1)$ respectively, which are shown in Figure 12.8. (Incidentally, the latter is an example of an *elliptic curve*, mentioned in Section 10.4.)

As the degree of the equation increases, more and more fancy shapes are possible. In fact, any shape you can draw can be approximated by an **algebraic curve**, that is, a curve defined by a polynomial in x and y. Figure 12.9 shows an example: a figure-eight curve defined by the fourth-degree equation

$$(x^2 + y^2)^2 = x^2 - y^2.$$

This curve is known as a **lemniscate**, from the Greek word for "ribbon". It was known to the Greeks (though they did not know its equation) as a

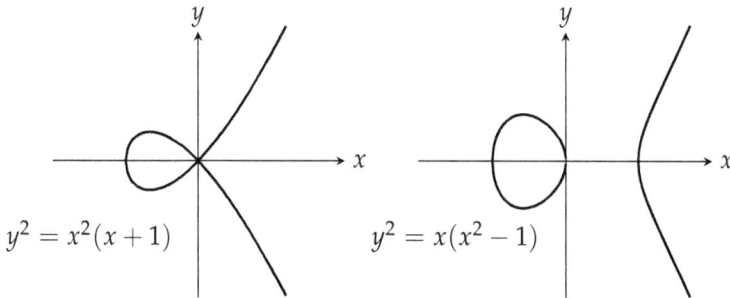

Fig. 12.8   Other cubic curves

section of the **torus**, the bagel-shaped surface generated by moving a small circle around a large circle (Figure 12.10).

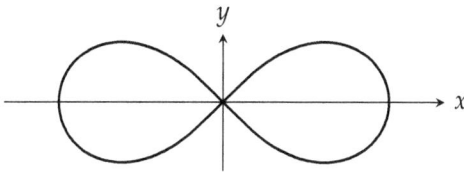

Fig. 12.9   The lemniscate of Bernoulli

This lemniscate is now named after Jakob Bernoulli, who investigated its arc length in 1694. It has an important place in the history of calculus as well as in the history of algebraic geometry. Finally, Figure 12.10 shows the lemniscate as it was first envisaged by the Greek mathematician Perseus (not the Perseus who slew Medusa) around 150 BCE.

Fig. 12.10   The lemniscate as a section of the torus

12.4   Tangents to Algebraic Curves

Before we see what algebraic geometry has to say about tangents it should be stressed that Apollonius knew all about the tangents to conic sections. Among his many propositions about tangents are constructions of tangents to parabolas, ellipses, and hyperbolas at any given point (Apollonius *Conics*, Book I, Propositions 33–35, which may be found in Heath (1896), pp. 25–27). Thus, again, to see the benefits of the algebraic approach we need to look at curves of higher degree.

Fermat was not only a pioneer of algebraic geometry, but also a pioneer of calculus. As every calculus student knows, one of the first problems in calculus is finding tangents to curves. Fermat solved this problem by a method that works for all algebraic curves; in fact, it shows that there is a purely algebraic method of finding tangents to algebraic curves. Finding these tangents is easier by calculus, but in theory it can be done by algebra, and it brings to light the important concept of **multiplicity** in algebraic geometry.

As we learned in Section 4.3, Diophantus found the far from obvious rational point $(21/4, 71/8)$ on the cubic curve $y^2 = x^3 - 3x^2 + 3x + 1$ by substituting $y = 3x/2 + 1$. The latter equation happens to represent the *tangent* to the curve at its "obvious" rational point $(0, 1)$, but I haven't yet explained *why* it is a tangent. The answer is a mixture of geometry and algebra.

Substituting $y = 3x/2 + 1$ in $y^2 = x^3 - 3x^2 + 3x + 1$ gives the equation

$$x^3 - 3x^2 + 3x + 1 = \left(\frac{3x}{2} + 1\right)^2$$

$$= \frac{9x^2}{4} + 3x + 1.$$

Subtracting $3x + 1$ from each side then gives the easy cubic equation

$$0 = x^3 - \frac{21x^2}{4} = x^2\left(x - \frac{21}{4}\right),$$

which has the solution $x = 21/4$. But it also has the solution $x = 0$, and it has this solution *twice*, in the sense that the factor x occurs twice. We can say, in turn, that the line $y = 3x/2$ "meets the curve twice" where $x = 0$. This is how algebra sees a tangent!

The geometric reason for saying that a tangent "meets the curve twice" is that the tangent at a point P on a curve is the limiting position of a line through P that *really* meets the curve twice (or more), as Figure 12.11 suggests. The line $\mathcal{L}$ through P close to the tangent $\mathcal{T}$ meets the curve at point P' close to P, and as $\mathcal{L}$ approaches $\mathcal{T}$ by rotating about P, P' approaches P. Algebraically, two different factors corresponding to P and P' become a repeated factor.

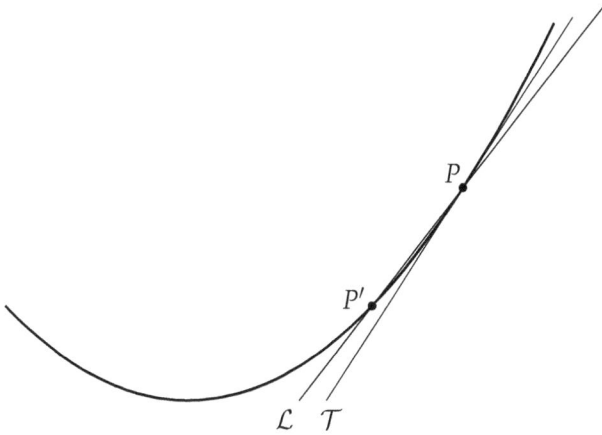

Fig. 12.11   A tangent and a nearby line

We can see this process in action with the parabola $y = x^2$, for which the x-axis $y = 0$ is an obvious tangent line at O. A nearby line through O has equation $y = \varepsilon x$ and it meets the parabola at a second point O' where $\varepsilon x = x^2$; namely, at $x = \varepsilon$ (Figure 12.12). As ε approaches zero, the two points O and O' come together.

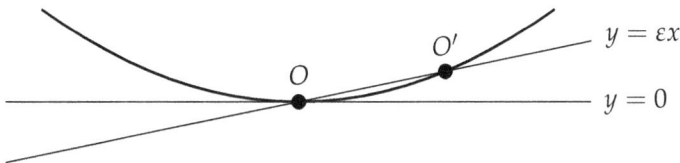

Fig. 12.12   Intersection of multiplicity 2 with parabola

The tangent $y = 0$ meets the parabola $y = x^2$ where $0 = x^2$, so the solution $x = 0$ is said to be of **multiplicity** 2, since the factor x occurs twice. It is also possible for a tangent to meet a curve with multiplicity 3 or more. For example, the line $y = 0$ meets the curve $y = x^3$ with multiplicity

3, since the equation $0 = x^3$ contains the factor x three times. And indeed it is geometrically clear that three points of intersection with a nearby line $y = \varepsilon x$ (Figure 12.13) come together as ε approaches zero. This behaviour, of course, is what you would expect at an inflection point.

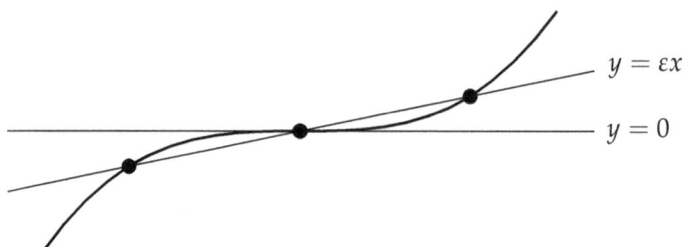

Fig. 12.13   Intersection of multiplicity 3

12.5   Rational and Integer Points on Curves

The Chord (or Secant) Method

Fermat discovered a method, possibly inspired by Diophantus, for finding rational solutions of quadratic equations in two unknowns. The method starts with an "obvious" solution and uses it to find *all* rational solutions. Since a quadratic equation represents a conic section, we can interpret the method as a way to find *rational points* on a conic section, where a rational point is one whose coordinates are both rational. The method is most naturally viewed geometrically, as I will show in the case of the unit circle,

$$x^2 + y^2 = 1.$$

Here it also gives a new way to find **Pythagorean triples**, previously mentioned in Section 3.7, where we derived Euclid's formula for them.

One obvious rational point on the unit circle is $P = (-1, 0)$, though any rational point will do. The method is simply to *draw a line of rational slope through the known rational point P and find its other intersection, Q, with the curve* (Figure 12.14). For any rational point Q the slope of the line PQ is certainly rational, so the method will certainly find all rational points. Conversely, any point Q hit by a line of rational slope through P will be rational, for algebraic reasons we will see shortly.

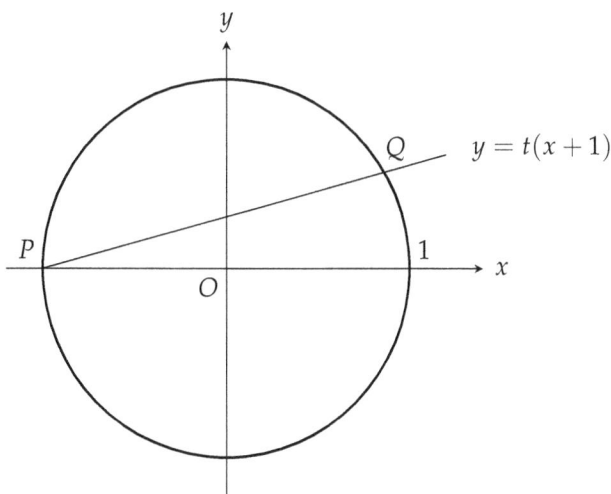

Fig. 12.14   Chord construction of rational points

The line of slope t through $P = (-1,0)$ has equation $y = t(x+1)$, so it meets the circle $x^2 + y^2 = 1$ where

$$1 = x^2 + t^2(x+1)^2$$
$$= x^2 + t^2(x^2 + 2x + 1)$$
$$= (1+t^2)x^2 + 2t^2x + t^2$$

and therefore

$$(1+t^2)x^2 + 2t^2x + t^2 - 1 = 0.$$

Now I could go ahead blindly and solve this equation by the quadratic formula, but it is more enlightening to first divide through by $1 + t^2$, which gives

$$x^2 + \frac{2t^2}{1+t^2}x + \frac{t^2-1}{1+t^2} = 0.$$

Then, since $x = -1$ is a solution of this equation (corresponding to the point P), the left-hand side must have a factor $(x+1)$. It follows from the constant term that the *other* factor must be

$$\left(x + \frac{t^2-1}{1+t^2}\right),$$

in which case the other solution, corresponding to the point Q, is $x = (1-t^2)/(1+t^2)$.

Thus the x-coordinate of the point Q is $(1 - t^2)/(1 + t^2)$, which is rational when t is. The y-coordinate is found from the equation $y = t(x + 1)$ of the line PQ to be

$$t \left(\frac{1 - t^2}{1 + t^2} + 1 \right) = t \frac{1 - t^2 + 1 + t^2}{1 + t^2} = \frac{2t}{1 + t^2},$$

which is also rational when t is. Thus each and every rational point on the unit circle, other than $(-1, 0)$, has the form

$$(x, y) = \left(\frac{1 - t^2}{1 + t^2}, \frac{2t}{1 + t^2} \right), \quad \text{for some rational } t.$$

If we write $t = q/p$, where p and q are integers, then we also find each rational point $(x, y) \neq (-1, 0)$ has the form

$$(x, y) = \left(\frac{p^2 - q^2}{p^2 + q^2}, \frac{2pq}{p^2 + q^2} \right), \quad \text{for some integers } p, q. \tag{*}$$

What does this have to do with Pythagorean triples? Well, a Pythagorean triple (a, b, c) satisfies $a^2 + b^2 = c^2$, and hence also

$$\frac{a^2}{c^2} + \frac{b^2}{c^2} = 1.$$

This says that $(a/c, b/c)$ is a rational point on the unit circle, and therefore, by the formula (*) for rational points,

$$\frac{a}{c} = \frac{p^2 - q^2}{p^2 + q^2}, \quad \frac{b}{c} = \frac{2pq}{p^2 + q^2} \quad \text{for some integers } p, q.$$

This finally gives

$$a = (p^2 - q^2)r, \quad b = 2pqr, \quad c = (p^2 + q^2)r \quad \text{for some integers } p, q, r,$$

which are Euclid's formulas.

The geometric method just described is perhaps not much simpler than the algebraic one described in Section 3.7. However, it has the advantage of showing that Pythagorean triples are very thick on the ground. More precisely, rational points are very thick on the circle – as thick as the rational numbers are on the line – because we hit one for every line of rational slope through the point P.

Other Conic Sections

The method just described will work for any conic section that has a rational point, as long as the equation of the conic has rational coefficients. In particular we can find all the rational points on hyperbolas of the form $x^2 - Ny^2 = 1$, for any natural number N, because they all have the rational point $x = 1, y = 0$.

I mention these because they include **Pell equations** (Section 3.5), where we seek *integer* solutions, and these are not nearly so easy to find. In fact, when Fermat in 1657 posed the solution of Pell equations to the English mathematicians, the first solutions he got (from Wallis and Brouncker) were the rational solutions. Fermat thought that these were trivial, so he had to explain that he was looking for integer solutions. Wallis and Brouncker were able to find these too, though it took them a few months. We pick up this story again in Section 13.5, and it may be found in Weil (1984), p. 92.

Later work on Diophantine equations led to the realisation that more sophisticated algebraic methods were needed to properly understand the problems of finding integer solutions. For finding rational solutions, however, geometric tricks are available, at least for cubic curves.

The Chord Method on a Cubic Curve

The chord method does not generally work on cubic curves, but it can if the curve intersects itself at a rational point. To illustrate, let's look at another famous cubic curve: $x^3 + y^3 = 3xy$, which is called the **folium of Descartes**. It gets its name (the Latin word for "leaf") from the leaf-shaped portion in the positive quadrant. Descartes, who had a surprisingly traditional view of geometry, looked only at the positive quadrant because he accepted only positive lengths. Thus his coordinates were only the positive numbers.

Figure 12.15 includes the parts with negative coordinates, showing clearly how the folium crosses itself at the point $(0,0)$. Now a line $y = tx$ through this "double point" O meets the curve $x^3 + y^3 = 3xy$ in another point P whose coordinates satisfy

$$x^3 + t^3 x^3 = 3tx^2.$$

This equation has the double solution $x = 0$, corresponding to the double point O, but also the solution

$$x = \frac{3t}{1 + t^3},$$

obtained by cancelling x^2. Since this point P is on the line $y = tx$, its y coordinate is $3t^2/(1 + t^3)$. Both the x and y coordinates are rational when t is, so a line of rational slope t through O meets the curve at the rational point

$$P = \left(\frac{3t}{1 + t^3}, \frac{3t^2}{1 + t^3} \right).$$

Conversely, the line from O to any rational point P has rational slope, so the chord construction has given us *all* the rational points on the folium.

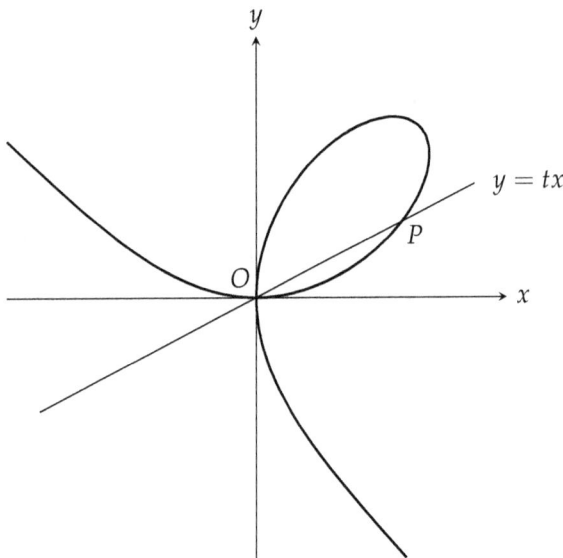

Fig. 12.15   The folium of Descartes

The folium was introduced by Descartes in 1638, as a challenge to Fermat. Descartes challenged Fermat to find the tangent at an arbitrary point of the folium – hoping to stump him – but Fermat found a solution easily. As mentioned earlier, Fermat was also a pioneer of calculus, and his method for finding tangents to algebraic curves was an early form of differentiation.

Even though Fermat knew his way around the folium, algebraically, we don't know whether he ever drew a picture of it. Rather surprisingly, the first known drawing that shows the curve going to infinity, together with its asymptote, was given in a letter by the Dutch scientist Christiaan

Huygens in 1692. Huygens' rather shaky effort is shown in Figure 12.16. The original may be seen at

https://gallica.bnf.fr/ark:/12148/bpt6k77858m/f357

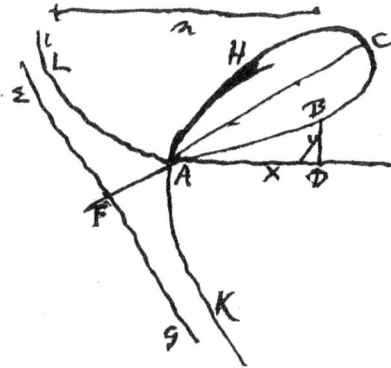

Fig. 12.16 Huygens' drawing of the folium

12.6 The Tangent Method

The equation of Diophantus from Section 4.3, $y^2 = x^3 - 3x^2 + 3x + 1$, for which the tangent at the "obvious" rational point $(0,1)$ leads to the much less obvious one, $(21/4, 71/8)$, illustrates a general method for finding rational points on cubic curves. This **tangent method** applies to any cubic equation with rational coefficients. Provided there is at least one rational point P on the curve, the tangent at P will meet the curve at another rational point, Q. Briefly, the reason is this:

Suppose $P = (a,b)$ is a rational point on the curve given by a cubic equation $p(x,y) = 0$. The tangent at P will be of the form $y = rx + s$, with r and s rational if the coefficients of $p(x,y)$ are rational. Then if we substitute $rx + s$ for y in the equation for the curve we get the cubic equation

$$p(x, rx + s) = 0.$$

This equation must have a double root $x = a$, because of the double contact of the tangent at the rational point $P = (a,b)$, which means that

$$p(x, rx + s) = (x - a)^2(cx + d) \quad \text{for some numbers } c, d.$$

The factor $cx + d$ on the right-hand side is the quotient of $p(x, rx + s)$, which has rational coefficients, by its factor $(x - a)^2$, so c and d are also rational. This gives $x = -d/c$ as the x coordinate of the other intersection, Q, whose y coordinate $rx + s$ is also rational. Thus Q is a rational point.

This shows how one rational point leads to another.

However, compared with the chord method for conic sections, and for a few cubic curves like the folium, the tangent method gives a rather meagre supply of rational points. If we are lucky, the tangent at the new rational point Q will give a rational point R different from P and Q, and so on, so we might get infinitely many rational points this way. Alternatively, for any *two* rational points P and Q we can get a third by drawing the chord through P and Q, since a cubic equation with rational coefficients, and two rational roots, has its third root rational too. So drawing chords is another option.

It seems that the first mathematician to fully grasp the role of chords and tangents in finding rational points was Newton, in unpublished notes from the early 1670s. He made these notes not long after the first publication of Fermat's work, so Newton evidently took up the baton from Fermat.

In Fermat's and Newton's day it was not at all clear that the chord and tangent methods, together, could manage to produce all the rational points on a cubic curve. It was not until the 1920s that Louis Mordell was able to prove that, for each cubic curve, *finitely many* rational points suffice to produce all the others. But we still don't have a general method for finding the magic finite set that produces all its rational points on a given curve. Cubic curves are difficult!

Chapter 13

Fermat's Number Theory

13.1 Introduction

Number theory concerns the natural numbers $0, 1, 2, 3, \ldots$, addition and multiplication. You may say, what about subtraction and division – not to mention exponentiation and fancier functions such as the nth prime number? Well, if you think about it, all the latter functions *arise from* addition and multiplication and can be defined in terms of them. For example, the **difference** $a - b$ can be defined as the number c such that $a = b + c$. Of course, taking differences may take us outside the natural numbers into the **integers**; for example $3 - 7 = -4$. Similarly, we can define the **quotient** a/b to be the number c such that $a = b \times c$, and taking quotients may take us outside the integers into the fractional, or **rational** numbers. And you can probably think of other operations that take us further than that.

So, before I start on Fermat's work, I'd like to expand on the concept of number theory. This involves numbers of various types and also the properties of addition and multiplication.

First, the system of integers, $\mathbb{Z}$. This includes both positive and negative numbers and zero, which are subject to addition, subtraction, and multiplication – but not generally to division. In fact, no number system allows unlimited division, since division by zero is not possible, but the rational number system, $\mathbb{Q}$, allows division by all its nonzero members. In these systems, subtraction can be "reversed" by addition, and division (by a nonzero number) can be "reversed" by multiplication. The basic properties of addition and multiplication are made precise in the concept of **field**, defined in Section 9.6. But to see that addition and multiplication are not as simple as they look, it's enough to look at some typical questions of number theory.

We've already seen some of them: for example, finding positive integers x, y, z such that $x^2 + y^2 = z^2$ (Pythagorean triples) or such that $x^2 - 61y^2 = 1$ (Bhaskara II's Pell equation). Also, finding perfect numbers, or Mersenne primes (Section 3.6). More generally, finding integer or rational solutions to Diophantine equations has been a source of interesting problems in number theory since ancient times and, as we have seen, it arose not only in Greece, with Diophantus, but also in the mathematics of India and China. Here is a small selection:

$$ax + by = c, \quad \text{(the Euclidean algorithm deals with this one)}$$
$$a^3 + b^3 = c^3 + d^3,$$
$$\text{and} \quad x^n + y^n = z^n. \quad \text{(Fermat's last theorem)}$$

The solutions of the second equation have become known as the taxicab numbers (Wikipedia, Taxicab number). These were named after a conversation between Hardy and Ramanujan, when Hardy was visiting Ramanujan in hospital. As Hardy (1940), p. 12, himself recalled

> I remember going to see him once when he was lying ill in Putney. I had ridden in taxi-cab No. 1729, and remarked that the number seemed to me rather a dull one, and that I hoped that it was not an unfavourable omen. "No," he replied, "it is a very interesting number; it is the smallest number expressible as a sum of two cubes in two different ways."

At this stage I can only say that mathematicians often have conversations that normal people would consider at best "quaint". They would probably yawn, if they heard mathematicians trying to list all the interesting numbers. The list might start with 1 as the first interesting number because it is the smallest positive integer; 2 could be included since it is the only even prime; 3 is the smallest odd prime; 4 is the smallest even square; and so on. I challenge you to find all of the interesting numbers up to 1729! And what do you think is the first *un*interesting number? I'm sure that you have nothing better to think about.

The third equation, for $n > 2$, has no solution for positive integers x, y, z, as was mentioned in connection with Fermat's last theorem in Section 10.4. Its proof was completed only in 1994, but Fermat himself found a proof for $n = 4$, which we'll see in Section 13.6.

13.2 Amicable Numbers and More

There are two points to this section. The first is to get another glimpse at what pure mathematicians do and why, and the second is to point out a site that has a large number of sequences/patterns that might be useful to you.

A Variant of Perfect Numbers

Now way back in this book I talked about perfect numbers (Section 3.6). To save you looking back, a perfect number is one whose proper divisors add to the number itself. So, the proper divisors of 6 are 1, 2, 3 (6 itself is excluded) and $1 + 2 + 3 = 6$. **Amicable numbers** are pairs of numbers where the sum of the proper divisors of one number equals the other number and vice versa. The smallest amicable pair is (220, 284), and it is amicable because the proper divisors of 220 are 1, 2, 4, 5, 10, 11, 20, 22, 44, 55, and 110 and their sum is 284; the proper divisors of 284 are 1, 2, 4, 71, and 142 and their sum is 220. You can think of a perfect number as being a self-amicable number.

In 1636, Fermat produced an amicable pair (17296, 18416). A couple of years later, Descartes found this one: (9363584, 9437056). (When you have nothing else to do you might like to check these pairs. I suggest that you write a program to do this.) Fermat's and Descartes' work was fine, but they didn't know that, centuries earlier, someone from Baghdad had already found those pairs.

Historically, Pythagoras, or one of the Pythagoreans, kicked off the amicable number idea. It was they who found the (220, 284), pair.

Then Thabit ibn Qurra (circa 830–901), the Baghdadian above, managed to produce a formula for getting more pairs. This is what he did. Let

$$p = 3 \times 2^{n-1} - 1$$
$$q = 3 \times 2^n - 1, \text{ and}$$
$$r = 9 \times 2^{2n-1} - 1.$$

Now there's a bit more. First, $n > 1$, and second, p, q and r have to be primes. Given all of that, $2^n pq$ and $2^n r$ are a pair of amicable numbers. Using this method, Thabit managed to find the amicable pairs (220, 284), (17296, 18416), (9363584, 9437056). Later, Euler (see Wikipedia, Amicable numbers) managed to extend this result and soon found 58 more new amicable pairs. The number of amicables had grown to 390 by 1946. Then computers started to get to work on the problem and the number of amicable pairs started to rocket. The present number of known amicable pairs is 1,227,820,104.

Why bother? Well, what mathematicians would like to find for any problem is closure in some sense. In this situation, they would like to know all amicable numbers. Failing that, a formula for some might be useful. At least it would be nice to know that there were an infinite number of amicable pairs. In looking at these, ideas might come up, leading to interesting theorems that have wider application.

But hang on a minute. Thabit had a formula. So surely we have a way to find an infinite number of amicables. Well, it's not that easy. The difficulty is that p, q and r have to be primes and there is no telling which n produce that extra condition. Are there only a finite number of amicables that you can get from Thabit's formula? Who knows? But we'd all like to know that.

For the moment all mathematicians can do is to find more and more to see if there is a pattern. You'll see the same thing with Fermat's last theorem (Section 13.6). More and more cases were found for specific numbers until Wiles came up with a proof that was far from anything Fermat knew about.

The Online Encyclopedia of Integer Sequences

At this point I want to introduce you to the Online Encyclopedia of Integer Sequences, OEIS, site: oeis.org. If you happen to be interested in a given sequence you can find out more about it by typing in the first 6 or so members of the sequence. When I typed in 1, 1, 2, 3, 5, 8 it said, among other things, A000045: Fibonacci numbers:

$$F(n) = F(n-1) + F(n-2) \quad \text{with} \quad F(0) = 0 \quad \text{and} \quad F(1) = 1$$

That is, the number given to the sequence, its name and a recurrence relation that defines it. The numbers go from 0 up to 102334155: at the end you can look up a list of the numbers, a graph, references, history, text, internal format.

And then more sequence numbers where these sequences start with the same numbers as the present one does. And the whole list above is repeated for the new sequence.

For the Fibonacci sequence there is enough written on the OEIS to write a project or a whole book.

Amicable number pairs appear under three OEIS numbers. If you enter increasing pairs in order, that is 220, 284, 1184, 1210, 2620, 2924, you'll find the sequence A063990. A002025 and A002046 are the smallest and biggest numbers, respectively, of amicable pairs.

Fermat shows up again with A019434, the Fermat primes. Fermat's *numbers* are of the form

$$\mathcal{F}(n) = 2^{2^n} + 1.$$

If $n = 0$, I get $\mathcal{F}(0) = 3$; if $n = 1$, I get $\mathcal{F}(1) = 5$; if $n = 2$, $\mathcal{F}(2) = 17$. It's beginning to look as if I should have called Fermat numbers the Fermat primes. The primes continue: $\mathcal{F}(3) = 257$ and $\mathcal{F}(4) = 65537$ are primes. In a letter to Pascal, Fermat claimed that all the $\mathcal{F}(n)$ are primes. Well, you might like to test $\mathcal{F}(5) = 4294967297$. Euler discovered that $\mathcal{F}(5)$ is *not* a prime, and the current thinking is that there are no more Fermat primes after $\mathcal{F}(4)$. This gets the attention of people who like numbers, but there is also an *entirely different* reason to be interested in the Fermat primes: they solve the ancient Greek problem of finding the regular polygons constructible by ruler and compass! (For this problem, see Section 2.5 to refresh your memory.)

As mentioned in Section 2.5, it was proved, basically by Gauss, that a regular n-gon is constructible if and only if n is the product of a power of 2 and distinct Fermat primes. In particular, the regular 17-gon is constructible (Gauss discovered this when he was 19 years old), and so are the regular 257-gon and the regular 65537-gon. But we won't know all of the constructible n-gons until we know all the Fermat primes!

13.3   Fermat's Little Theorem

Although the word "little" is attached to the title of this result, it turns out to be a basic result in number theory. The theorem is called "little" only to distinguish it from Fermat's *last* theorem, which is sometimes called his "great" theorem.

One way to state Fermat's little theorem is: *if p is prime then p divides $n^p - n$, for any whole number n*. This is a raw form of the theorem, involving concepts familiar to Fermat. Later we'll see that the theorem becomes more refined and elegant when expressed in more modern language – especially the language of **congruence** – but for now I'll approach the theorem as Fermat might have. I will assume that Fermat knew something about the binomial coefficients, as he must have through his acquaintance with Pascal, if not before.

Let's begin by taking another look at Pascal's triangle from Section 11.5 (Figure 13.1), where we can notice an interesting divisibility phenomenon. Look especially at the prime-numbered rows:

$$1$$
$$1 \quad 1$$
$$1 \quad 2 \quad 1$$
$$1 \quad 3 \quad 3 \quad 1$$
$$1 \quad 4 \quad 6 \quad 4 \quad 1$$
$$1 \quad 5 \quad 10 \quad 10 \quad 5 \quad 1$$
$$1 \quad 6 \quad 15 \quad 20 \quad 15 \quad 6 \quad 1$$
$$1 \quad 7 \quad 21 \quad 35 \quad 35 \quad 21 \quad 7 \quad 1$$
$$\vdots$$

Fig. 13.1   Divisibility in Pascal's triangle

row 2, which is   1 2 1

row 3, which is   1 3 3 1

row 5, which is   1 5 10 10 5 1

row 7, which is   1 7 21 35 35 21 7 1

Each row p begins and ends with 1, but the binomial coefficients $\binom{p}{n}$ in between are all divisible by the row number p. Why is this so?

Well, Pascal's formula for the binomial coefficients says

$$\binom{p}{n} = \frac{p(p-1)(p-2)\cdots(p-n+1)}{n!},$$

which we notice has a factor p in the top line but not in the bottom line, since $n < p$. Since each $\binom{p}{n}$ is a whole number, its bottom line $n!$ must divide the product of factors in the top line, *not including the factor* p, since p is prime and hence is not divisible by $n, n-1, n-2, \ldots$. It follows that p is a factor in the prime factorisation of $\binom{p}{n}$ when p is prime.

Now we can see how divisibility by p also occurs in $n^p - n$, for any n. First, let's take $n = 2 = 1 + 1$, so

$$n^p = (1+1)^p = 1 + \binom{p}{1}1^1 + \binom{p}{2}1^2 + \cdots + \binom{p}{p-1}1^{p-1} + 1^p$$

by the binomial theorem

$$= 2 + \binom{p}{1}1^1 + \binom{p}{2}1^2 + \cdots + \binom{p}{p-1}1^{p-1}$$

so, since $2 = n$,

$$n^p - n = \binom{p}{1}1^1 + \binom{p}{2}1^2 + \cdots + \binom{p}{p-1}1^{p-1}. \qquad (*)$$

Consequently, p must divide each term on the right-hand side of (*), and hence it divides $n^p - n$ when $n = 2$. The ideas in this proof can be extended to any value of n by induction, but let's now look at Fermat's little theorem from a more modern point of view.

Fermat's Little Theorem via Congruences and Inverses

Fermat's little theorem, like many number theory questions since ancient times, involves the concept of divisibility. Around the year 1800, Carl Friedrich Gauss found a way to simplify many calculations and proofs involving divisibility, and he introduced it at the beginning of his book Gauss (1801), called the *Disquisitiones arithmeticae* (Arithmetical investigations). The key idea is that of **congruence**, which (like the concept of the same name in geometry) is something like equality, only a little looser.

For any integers a, b, and n we write

$$a \equiv b \pmod{n}$$

to mean that n divides $a - b$. In other words, a and b differ by a multiple of n. Informally, a and b are the same "up to multiples of n" or, as we also say, they are **congruent modulo** n. The formula $a \equiv b \pmod{n}$ is called a **congruence**, which we pronounce "a congruent to b, mod n".

What makes this concept useful, and good for calculation, is that congruences behave much the same as equations: they can be added and multiplied, as long as we keep the **modulus** n the same. In particular we have

$$\text{If } a_1 \equiv b_1 \pmod{n} \text{ and } a_2 \equiv b_2 \pmod{n}$$
$$\text{then } a_1 + a_2 \equiv b_1 + b_2 \pmod{n} \text{ and } a_1 \times a_2 \equiv b_1 \times b_2 \pmod{n}.$$

A famous example, called "casting out nines", is about congruence mod 9. Casting out nines is an old calculator's trick (mentioned in the *Liber abaci*) for finding the remainder when a number, say 8777, is divided by 9. The trick says that it is enough to divide the sum of the digits by 9. In this case $8 + 7 + 7 + 7 = 29$, which leaves remainder 2 when divided by 9. And 2 is indeed the remainder when 8777 is divided by 9. Why? Well,

$$8777 = 8 \times 10^3 + 7 \times 10^2 + 7 \times 10 + 7$$

and 10 is obviously congruent to 1, mod 9. But then 10^2 and 10^3 are also congruent to 1, mod 9, by multiplication of congruences. So we can convert the above equation to the congruence

$$8777 \equiv 8 \times 1 + 7 \times 1 + 7 \times 1 + 7 \pmod{9}$$
$$\equiv 8 + 7 + 7 + 7 \pmod{9}$$

since 10, 10^2, and 10^3 are "the same as" 1, mod 9. And this congruence says that 8+7+7+7 leaves the same remainder as 8777 when divided by 9.

You will notice that 10 is also congruent to 1, mod 3, so the same trick works for division by 3. In particular, a number is divisible by 3 if and only if the sum of its digits is divisible by 3.

To return to Fermat's little theorem: it deals with divisibility by a prime p, so the numbers n to think about are those that are "different" mod p, namely $0, 1, 2, \ldots, p - 1$. Because p is prime the numbers that are nonzero, mod p, have a remarkable property: they all have **multiplicative inverses**, mod p. Let's state this as a theorem:

Inverses mod p. For each a among $1, 2, \ldots, p - 1$ there is a b such that

$$a \times b \equiv 1 \;(\text{mod } p).$$

Proof. The meaning of $a \times b \equiv 1 \;(\text{mod } p)$ is that p divides $ab - 1$, which in turn means that $ab - 1 = mp$ for some integer m, or that

$$1 = ab + mp \quad \text{for some integer } m.$$

This looks like an instance of Bézout's identity that we discussed back in Section 3.2, and indeed it is. Since a is among $1, 2, \ldots, p - 1$, a is not divisible by p, and therefore $\gcd(a, p) = 1$. Bézout's identity then says that 1 can be written as $ab + mp$, for some integers b and m, so we have found the required b.   Q.E.D.

The existence of inverses was probably not known to Fermat in any form, but it gives a very elegant proof of his little theorem, particularly if we state the theorem in a slightly different way.

Fermat's little theorem. If a is a number not divisible by the prime p, then

$$a^{p-1} \equiv 1 \;(\text{mod } p).$$

Proof. Since a is not divisible by p we have $a \not\equiv 0 \;(\text{mod } p)$. Therefore, if we multiply a by $1, 2, \ldots, p - 1$ we get numbers

$$a, \quad 2a, \quad \ldots, \quad (p - 1)a,$$

which are all $\not\equiv 0 \;(\text{mod } p)$. This is because if $ab \equiv 0 \;(\text{mod } p)$ we get $b \equiv 0 \;(\text{mod } p)$ by multiplying both sides by the inverse of a.

In fact, $a, 2a, \ldots, (p - 1)a$ are the same numbers (mod p) as the numbers $1, 2, \ldots, p - 1$, except perhaps in a different order. First, they are distinct, because if we multiply each of $a, 2a, \ldots, (p - 1)a$ by the inverse of a we get back the distinct numbers $1, 2, \ldots, p - 1$. Then, since $a, 2a, \ldots, (p - 1)a$ are nonzero (mod p) they must be *all* of the nonzero numbers $1, 2, \ldots, p - 1$ (mod p).

It follows that

$$a \times 2a \times \cdots \times (p - 1)a \equiv 1 \times 2 \times \cdots \times (p - 1) \ (\text{mod } p).$$

Then we can cancel $1, 2, \ldots, p - 1$ from both sides by multiplying by their inverses, leaving

$$a^{p-1} \equiv 1 \ (\text{mod } p),$$

as required. Q.E.D.

This version of the little theorem looks a little different from Fermat's, even allowing for the fact that I have used the letter a instead of n. We can get back from this version to Fermat's in two easy steps:

(1) Multiplying both sides of $a^{p-1} \equiv 1 \ (\text{mod } p)$ by a gives $a^p \equiv a \ (\text{mod } p)$, which says that p divides $a^p - a$.
(2) We proved $a^{p-1} \equiv 1 \ (\text{mod } p)$ only for a not divisible by p. But if a is divisible by p then p obviously divides $a^p - a$.

13.4 The Legacy of Fermat's Little Theorem

Fermat's little theorem hints at the importance of inverses in mathematics, though this became clear only gradually. The next to exploit the idea was Leonhard Euler, who in 1758 gave a proof of Fermat's little theorem explicitly using inverses. In fact he found a more general theorem, generalising Fermat's little theorem from a prime modulus p to any modulus n.

Among the numbers $0, 1, 2, 3, \ldots, n - 1$ that are distinct mod n, Euler picks out the a that have multiplicative inverses. The number of these is given by a function $\varphi(n)$, now called the **Euler φ function**. For example, when $n = 8$ there are four invertible numbers, mod 8, namely 1, 3, 5, and 7, which are the numbers a among $1, 2, 3, 4, 5, 6, 7$ for which $\gcd(a, 8) = 1$. Thus $\varphi(8) = 4$.

Then a proof very similar to the above proof of Fermat's little theorem gives **Euler's theorem**:

For any invertible a, $a^{\varphi(n)} \equiv 1 \ (\text{mod } n)$.

In the 20th century, Euler's theorem made a surprise contribution to the world of espionage and e-commerce, in the form of a **cryptosystem** commonly known as RSA. (See Wikipedia, RSA.) Introduced in the 1970s, the system protects the security of internet transactions, ultimately by exploiting the fact that it is hard to factorise numbers with hundreds of digits. This is why I have from time to time mentioned the value of large prime numbers in this book. It turns out that Euler's theorem enables the encryption of a message, using the *product* of two large primes, which can feasibly be decrypted only by those who know the individual primes.

The system is called RSA after Rivest, Shamir, and Adleman, who published the first paper on the system in 1977. Much later, it was revealed that the system had in fact been discovered independently by Clifford Cocks in 1973 (see Wikipedia, Clifford Cocks). Cocks was then working for the British security service GCHQ, and his discovery remained secret until 1997. He still does not receive a lot of credit for RSA, though the British government honoured him with a knighthood in 2008.

13.5   Fermat and the Pell Equations

This book has mentioned Pell equations several times, most recently in Section 11.4, where it was pointed out how badly misnamed they are. Nevertheless, "Pell" is the name always used when we seek integer solutions x and y of the equation

$$x^2 - By^2 = 1,$$

where B is a positive square-free number. As we saw in Section 3.5, the $B = 2$ case had been studied by the Pythagoreans, who introduced a recurrence relation that found all solutions. Later Brahmagupta (Section 8.5) in the early 7th century, found a way to solve many cases of Pell's equation and even extended it to $x^2 - By^2 = k$, where $k = \pm 1, \pm 2$, and ± 4.

Later, in 1150, Bhaskara II extended Brahmagupta's work, as we saw in Section 11.1. Undoubtedly Fermat had stirred the Pell pot in Europe in the 17th century, making great strides forward; however, he had basically caught up with Bhaskara II and unknowingly had used his method, or something similar. We don't know exactly how Fermat solved Pell equations, but we do know that Brouncker used **continued fractions**, and that this became the favoured method in Europe, not only for solving Pell equations, but also for implementing the more basic process of the **Euclidean algorithm**.

So let's now look again at the Euclidean algorithm, and its relationship to continued fractions.

Continued Fractions

Brouncker's continued fraction for π, in Section 11.4, gives you an idea of what a continued fraction looks like, but in this section I will stick to a simpler kind. These are the **regular continued fractions**, and they have the form

$$a + \cfrac{1}{b + \cfrac{1}{c + \cfrac{1}{d + \cfrac{1}{\ddots}}}},$$

where $a, b, c, d, \ldots$ are positive integers and each numerator equals 1. They come in two flavours, finite and infinite, and both are related to the Euclidean algorithm (or its Indian equivalent, the "pulverisor"). In both cases, what "continues" is the process of division with remainder. Here are some examples.

First, let's look at the Euclidean algorithm on the pair of numbers 23, 4, using division with remainder.

If we divide 23 by 4 we get quotient 5 and remainder 3, so we replace 23 by its remainder, 3, and keep the divisor 4. We then have the pair 3, 4 so, dividing the larger by the smaller (namely 4 by 3) replaces 4 by its remainder 1, and we keep the divisor 3. We then have the pair 1, 3 and the next step will be the exact division (of 3 by 1), so the algorithm halts with the divisor 1 – which therefore equals $\gcd(23, 4)$.

That's pretty longwinded, but it can be concisely simulated by a calculation with fractions. The following table shows statements about division alongside equivalent equations between fractions, ending with a regular continued fraction.

Statement	Equation
23 divided by 4 gives quotient 5, remainder 3	$\frac{23}{4} = 5 + \frac{3}{4}$
Replace pair 3, 4 by 4, 3 (via $\frac{3}{4} = \frac{1}{4/3}$)	$= 5 + \frac{1}{4/3}$
4 divided by 3 gives quotient 1, remainder 1	$= 5 + \cfrac{1}{1 + \frac{1}{3}}$

Thus the original fraction 23/4 has been "expanded" in a way that reflects the running of the Euclidean algorithm on the pair 23, 4:

$$\frac{23}{4} = 5 + \cfrac{1}{1 + \frac{1}{3}}$$

The continued fraction on the right-hand side captures the whole run of the Euclidean algorithm, by displaying the whole sequence of quotients 5, 1, 3 (the last because 3 divided by 1 is 3). This will work with any initial fraction in lowest terms.

An example where the Euclidean algorithm runs a little longer is 53/37. In this case we get

$$
\begin{aligned}
\frac{53}{37} &= 1 + \frac{16}{37} \\
&= 1 + \frac{1}{37/16} \\
&= 1 + \frac{1}{2 + \frac{5}{16}} \\
&= 1 + \frac{1}{2 + \frac{1}{16/5}} \\
&= 1 + \frac{1}{2 + \frac{1}{3 + \frac{1}{5}}}.
\end{aligned}
$$

And this is where the algorithm, and the fraction, stops. As you can see, the continued fraction (the last line) displays the whole run of the Euclidean algorithm at a glance. In this case, it shows the successive quotients 1, 2, 3, 5. As with the Euclidean algorithm, the continued fraction process on the ratio of two whole numbers eventually stops, because it continually produces smaller whole numbers.

But nobody said you have to start with a ratio of whole numbers! In fact, the continued fraction process on an *irrational* ratio will run forever, and the results can be interesting. This version of the Euclidean algorithm is possibly as old as the finite version. Euclid used it as a way to test for irrationality. Here is an example: the ratio of $\sqrt{2} + 1$ to 1. The ratio of $\sqrt{2}$ to 1 will also give an infinite process but $\sqrt{2} + 1$ is a little neater, because of the fact that

$$
\sqrt{2} - 1 = \frac{1}{\sqrt{2} + 1}. \tag{*}
$$

This is convenient when you are continually inverting fractions.

Let's go! The process starts with the fact that $\sqrt{2} + 1$ divided by 1 equals 2 (the number of times you can subtract 1 before the remainder becomes less then 1), with remainder $\sqrt{2} - 1$. After that, the same thing happens over and over …

$$\sqrt{2} + 1 = 2 + \sqrt{2} - 1 \qquad \text{dividing } \sqrt{2} + 1 \text{ by } 1$$

$$= 2 + \frac{1}{\sqrt{2} + 1} \qquad \text{by (*)}$$

$$= 2 + \frac{1}{2 + \sqrt{2} - 1} \qquad \text{dividing } \sqrt{2} + 1 \text{ by } 1 \text{ again}$$

$$= 2 + \frac{1}{2 + \frac{1}{\sqrt{2}+1}} \qquad \text{by (*) again}$$

Clearly, it's 2s all the way down, so we can write

$$\sqrt{2} + 1 = 2 + \cfrac{1}{2 + \cfrac{1}{2 + \cfrac{1}{2 + \cfrac{1}{\ddots}}}}$$

This is an example of an **infinite continued fraction**. Also, it's a **periodic** continued fraction, in the sense that the sequence of quotients repeats. We get an infinite continued fraction for $\sqrt{2}$ simply by subtracting 1 from both sides, namely:

$$\sqrt{2} = 1 + \cfrac{1}{2 + \cfrac{1}{2 + \cfrac{1}{2 + \cfrac{1}{\ddots}}}}.$$

This type of continued fraction is called **ultimately** periodic, meaning that it becomes periodic after a certain point (here, after the first quotient 1).

It is fairly easy to prove that any ultimately periodic continued fraction is of the form $a + b\sqrt{c}$ for some rational numbers a, b, and c. A famous example is the simplest infinite continued fraction

$$\varphi = 1 + \cfrac{1}{1 + \cfrac{1}{1 + \cfrac{1}{1 + \cfrac{1}{\ddots}}}}.$$

Notice that the continued fraction beneath the first numerator on the right-hand side is … φ itself! Therefore φ satisfies the equation

$$\varphi = 1 + \frac{1}{\varphi}, \quad \text{that is,} \quad \varphi^2 = \varphi + 1,$$

which is none other than the equation for the **golden ratio** $\varphi = (1+\sqrt{5})/2$ discussed in Section 9.3.

The other direction of the theorem about ultimately periodic continued fractions is also true but harder to prove. The hard part is showing that the continued fraction for $\sqrt{c}$ is ultimately periodic, for any positive nonsquare integer c. This was first proved by Joseph-Louis Lagrange in 1768. In the process he also proved that the Pell equation $x^2 - By^2 = 1$ always has a solution, when B is a nonsquare positive integer. A *connection* between Pell equations and continued fractions had been known ever since Brouncker used them to solve Pell equations. Brouncker's method depended on the fraction becoming periodic – which it always did – but no one knew why until they saw Lagrange's proof.

Every positive irrational number has an infinite regular continued fraction, but its denominators don't necessarily form an orderly sequence. For example, the regular continued fraction for π has the successive denominators

$$7, 15, 1, 292, 1, 1, 1, 2, 1, 3, 1, 12, 2, 1, 1, 2, \ldots$$

(The 7 at the beginning occurs because π is close to $3\frac{1}{7}$.) However, if we allow arbitrary numerators, then nicer formulas are possible, such as Brouncker's beauty from Section 11.4:

$$\frac{4}{\pi} = 1 + \cfrac{1^2}{2 + \cfrac{3^2}{2 + \cfrac{5^2}{2 + \cfrac{7^2}{2 + \cfrac{9^2}{2 + \ddots}}}}}$$

13.6   Fermat's Last Theorem (FLT)

Introduction

This fuss is what happens when you scribble on the pages of a book. You worry the mathematical world for the next 350-plus years! On the other hand, if you are happy to wait for another 350 years it'll probably be the sort of question that a senior high-school student will be able to knock off in under an hour.

Anyway, it was all the fault of Diophantus (see Section 4.3) and his book called *Arithmetica*. While reading a passage on Pythagorean triples, Fermat wrote down on the nearest page something that looked like this (when translated into English):

> It is impossible for a cube to be a sum of two cubes, a fourth power to be a sum of two fourth powers, or in general for any number that is a power greater than the second to be the sum of two like powers. I have discovered a truly remarkable proof [of this result], but this margin is too small to contain it.

Formally he was saying that

if n is an integer greater than 2, then

$x^n + y^n = z^n$, has no nonzero integer solution for x, y and z.

It is likely that he soon realised that he had no proof. This can be supported by the fact that Fermat was keen on sending questions to European mathematicians for them to solve, that is, questions whose answer or proof he knew. If he had solved the FLT, then it is more than likely that he would have sent the FLT out as a question.

Obviously Fermat knew about Pythagorean triples (see Section 3.7). So he certainly would have known many solutions for the $n = 2$ case, such as:

$$(3, 4, 5), \quad (5, 12, 13), \quad (8, 15, 17), \quad (11, 60, 61) \quad \text{and so on.}$$

Indeed, he knew the general formula for Pythagorean triples, and he found an impressive use for it, as we will see in a moment. But what did he have in mind that day that persuaded him that $n = 3$ would get no solutions and neither would $n = 4, 5, 6$, and so on forever? Or was this just a note to himself to think about it some more? Or did he see a method that he eventually gave up on because it had too many cases to handle?

FLT for $n = 4$

Fermat proved FLT for $n = 4$ by actually proving something stronger:

Fermat's fourth power theorem. The equation $x^4 - y^4 = z^2$ has no positive integer solution.

It follows from this theorem that $a^4 + b^4 = c^4$ has no positive integer solution either, because if it had, we would have $c^4 - b^4 = a^4$, which solves the equation above with $c = x$, $b = y$ and $a^2 = z$. So it suffices to prove that the equation $x^4 - y^4 = z^2$ is impossible. Fermat did this by showing that any positive integer solution gives a *smaller* positive integer solution, which is ruled out by "infinite descent".

Proof. Suppose that the equation $x^4 - y^4 = z^2$ has a positive integer solution. First let me note that if x and y have a common divisor, that number would also divide z. In that case I could divide the solutions by the factor and I would have *coprime* (aka relatively prime) solutions x, y, z. So I'll assume that the putative solutions are coprime.

Now I'll factor the left-hand side of the original equation to give

$$(x^2 + y^2)(x^2 - y^2) = z^2.$$

Because x and y are coprime, then there are two cases. Either

(i) one of x or y is even, in which case $x^2 + y^2$ and $x^2 - y^2$ are coprime, because any common divisor of $x^2 + y^2$ and $x^2 - y^2$ divides their sum $2x^2$ and their difference $2y^2$, and hence also x and y, or

(ii) x and y are both odd and so $x^2 + y^2$ and $x^2 - y^2$ have a greatest common divisor of 2.

First consider Case (i): As the two factors $x^2 + y^2$ and $x^2 - y^2$ have no common divisor, and their product equals z^2, these individual factors have to be squares. Let

$$x^2 + y^2 = s^2 \quad \text{and} \quad x^2 - y^2 = t^2 \tag{*}$$

Since exactly one of x, y is even, s and t are both odd and so their sum and difference are both even. Let $u = (s + t)/2$ and $v = (s - t)/2$. Then adding the two equations in (*), I get

$$\frac{s^2 + t^2}{2} = x^2, \quad \text{and from the definition of } u \text{ and } v \text{ I get}$$

$$u^2 + v^2 = \frac{s^2 + 2st + t^2}{4} + \frac{s^2 - 2st + t^2}{4} = \frac{s^2 + t^2}{2} = x^2 \tag{**}$$

This says (u, v, x) is a Pythagorean triple, so (by Euclid's formulas from Section 3.7) there are coprime numbers m and n such that

$$u = 2mn, \quad v = m^2 - n^2, \quad \text{and} \quad x = m^2 + n^2.$$

But, taking the difference of the equations (*), $y^2 = (s^2 - t^2)/2 = 2uv$. Since x and y are coprime, then so are u and v. Hence, without loss of generality, $u = 2d^2$ and $v = e^2$.

So $u = 2mn = 2d^2$. But since m and n are coprime, they are squares. If I put $m = g^2$ and $n = h^2$, then $v = m^2 - n^2 = (g^2)^2 - (h^2)^2 = e^2$. Thus

(g, h, e) is another solution to the original equation and *it is smaller than the previous one*. So I can keep repeating this process to get a smaller solution than (g, h, e), then an even smaller one and so on for ever. But this is what infinite descent is all about. Starting from a given set of numbers, there is only a *finite* set of numbers beneath them. So this contradiction shows that the original assumption is incorrect in Case (i).

Case (ii): This case goes through similar steps as I used in Case (i) and the infinite descent provides the contradiction. Q.E.D.

Fermat's proof (which is similar to the one above) of FLT for $n = 4$ is a fine example of his work, and interesting in the way it stands on the shoulders of Euclid and Diophantus. The ultimate proof of FLT by Andrew Wiles in 1994 used the much more sophisticated theory of **elliptic curves**, which is too advanced to be even sketched in this book. But, remarkably, Wiles took the baton from Fermat himself because Fermat's investigations of cubic curves, mentioned in Section 12.6, were among the first results in the theory of elliptic curves.

Before Wiles' proof, however, even the unsuccessful attempts to prove FLT were fruitful, because they led to advances in mathematics. One such advance was made by Sophie Germain, whom we discuss in the next section.

13.7 Sophie Germain

Early Life

I'm sorry but this isn't one of the nice standard biographies that I have written for the mathematicians in other chapters of this book. The problem is that the subject this time has a "Marie" and a "Sophie" for her first name. And that made a lot of difference in late 18th and early 19th century France.

Marie-Sophie Germain (Figure 13.2) was born in Paris on the first of April 1776. Her father, Ambroise-Francois Germain was a wealthy merchant, as well as a goldsmith and a jeweller. Her mother was Marie-Madeleine and Marie-Sophie had two sisters. They were, in age order, Marie-Madeleine (born 1770), and Angelique-Ambroise (born 1779). Marie-Sophie became generally known as Sophie.

She was 13 years old when a constitutional monarchy was established in France, but the seizure of the Bastille quickly followed. The Revolution had begun; the monarchy had ended. There were two main effects on

Fig. 13.2   Sophie Germain (from Wikipedia, Sophie Germain)

the Germain family. First Ambroise-Francois took a high position in the government of Paris and went on to become a member of the National Assembly. Sophie's father decided that Paris wasn't a place for his daughters to be seen outdoors, so they were kept indoors.

The second effect was that, as a result of the first effect, Germain decided to read her way through her father's well-stocked library. In the library was *Histoire des mathématiques* by Étienne Montucla. Germain came across the story of Archimedes who had to be told by his servant when to eat. And further that Archimedes was killed by a Roman soldier because he wouldn't stop doing maths. (Maybe there was some autism there.) Of course, she learnt a lot of mathematics. As a result, she did what most of us would do: she decided to become a mathematician!

This decision led her to read maths books that her father had conveniently bought at some time. These included a book on arithmetic and one on calculus. In order to read books by foreign greats such as Newton and Euler she taught herself Latin and Greek. Her parents had no trouble in seeing these activities as bad for girls, so they took away her fire, her light and even her clothes so that she couldn't read at night. Germain's counter-attack was to wrap herself in blankets and read by the light of a candle. She was eventually caught one morning, asleep in the library and with her ink frozen by the cold air!

Germain and Lagrange

Germain turned eighteen in time to attend the École Polytechnique. At least she would have, except that girls didn't do such things then. However, the École had a policy of making lecture notes available for anyone who asked. As a result she "took" analysis lectures from the famous French mathematician Lagrange. She even pushed her luck and sent notes to him. The work that she sent Lagrange really impressed him and so he sought the person who had written it. Perhaps surprisingly for the time, Lagrange wasn't fazed by having a woman for a correspondent. Through Lagrange, Germain was introduced to other mathematicians. But the interactions were random and as a result she didn't get a thorough mathematical training, so she missed out on many fundamental processes and methods. She was also unhappy about her treatment by some mathematicians like Jerome Lalande, who suggested she should read a book on astronomy for women that had no maths in it at all.

But not all her experiences were bad for Germain. Germain obviously read Adrien-Marie Legendre's *Essay on the Theory of Numbers*, and their subsequent interactions were essentially collaborations on number theory. Some of Germain's work with Legendre is to be found in the second edition of his book, though he may not have recognised her contributions.

Germain and Gauss

Germain and the German mathematician Carl Friedrich Gauss also had a period of collaboration inspired by Gauss' *Disquisitiones arithmeticae*. The fact that Germain was a woman was not known to Gauss until the French army under Napoleon captured Braunschweig. Germain was worried that Gauss might meet the same fate as Archimedes. So she asked a French general known to her father to protect Gauss. As a consequence Germain decided that she had to tell him that she was a woman.

> In describing the honourable mission I charged him with, M. Pernety informed me that he made my name known to you. This leads me to confess that I am not as completely unknown to you as you might believe, but that fearing the ridicule attached to a female scientist, I have previously taken the name of M. LeBlanc in communicating to you those notes that, no doubt, do not deserve the indulgence with which you have responded.
>
> Sophie Germain, Letter to Gauss (20 February, 1807)

Elasticity

Germain became very interested in elasticity when she attended a visit to Paris by the German physicist Ernst Chladni. There she saw his experiments on vibrating plates that were sprinkled with sand. These experiments produced what are now called Chladni figures (see Figure 13.3). But Chladni was not able to explain their interesting symmetric shapes. As a result, the Institute de France established a competition to formulate a mathematical theory of elastic surfaces and indicate just how it agrees with empirical evidence.

On the due date two years later, Germain sent in her submission. She was the only entrant, as more established mathematicians had been put off when Lagrange let it be known that the necessary maths didn't yet exist. Germain didn't get an award because she hadn't based her work on physics and her maths knowledge was inadequate.

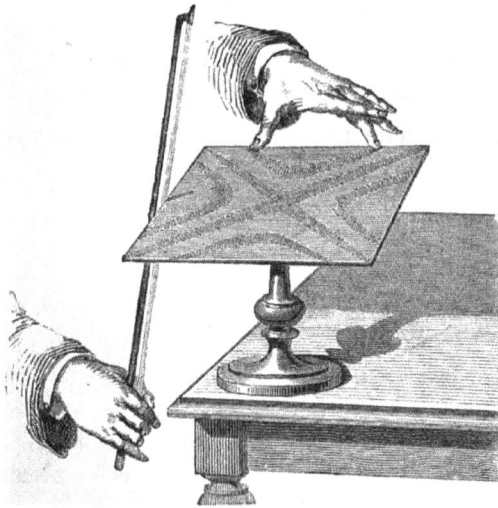

Fig. 13.3   Creating Chladni figures (from Wikipedia, Ernst Chladni)

Two years later at the next deadline, Germain was again the only person who submitted a solution. This time she could show that Lagrange's work was able to provide many cases of Chladni figures. But she could still not use empirical evidence. This time though she was given an honourable mention.

In 1815, Germain submitted her third attempt. Although her maths wasn't quite accurate, she was awarded a gold medal weighing a kilogram! (Worth about US $80,000 today.) But Germain didn't attend the prize ceremony, because she felt that the judges of the award hadn't really appreciated what she had done and she had not got the respect from the scientific community that she felt she deserved. See Bucciarelli and Dworsky (1980).

FLT

By 1819, many mathematicians had tried to solve FLT, but only a few specific cases, such as Fermat's $n = 4$, had been established as having no solutions.

Germain sent a letter to Gauss on 12 May, 1819, saying that

> Although I have laboured for some time on the theory of vibrating surfaces (to which I have much to add if I had the satisfaction of making some experiments on cylindrical surfaces I have in mind), I have never ceased to think of the theory of numbers ... A long time before our Academy proposed as the subject of a prize the proof of the impossibility of Fermat's equation, this challenge ... has often tormented me.

She then tried to find a grand plan that would prove FLT for an infinite number of values of n. In doing so, she was the first person to set out a programme for dealing with an infinite number of cases. For the details of her plan, and its evolution as she worked on it, see

https://www.math.mcgill.ca/darmon/courses/
12-13/nt/projects/Colleen-Alkalay-Houlihan.pdf

Bear in mind that the only values of n for which FLT has to be proved are $n = p$ an odd prime. So in what follows, p is an odd prime. Germain's ultimate plan broke FLT into two parts: showing that $x^p + y^p = z^p$ has no integer solutions for which either

(i) none of x, y, and z is divisible by p; or
(ii) exactly one of x, y, z is divisible by p.

Working from Germain's plan in 1823, Legendre found FLT is true for $p = 5$, and Germain independently found more cases. See Del Centina (2008), and Laubenbacher and Pengelley (2010). Germain and Legendre's work was extendible to $p < 197$. However, for bigger primes p the calculations required were beyond their calculating ability.

Working on Germain's two conditions became a common approach to finding more p for which FLT could be proved. And even toward the end of the 20th century, E. Grosswald wrote in *Mathematical Reviews*, reviewing Heath-Brown's paper *The first case of Fermat's last theorem*,

> The tools used, or quoted, in the proof are rather formidable and comprise, among others, a ... theorem of Faltings, old theorems of Sophie Germain and of Wieferich and Mirimanoff, and a generalization of Sophie Germain's theorem ...

Germain Summary

Germain developed breast cancer in 1829 and died two years later. She is buried in the Cimitiere du Pére Lachaise, Paris. There is a simple headstone marking the grave.

There is much more that I ought to have included here. For example, along with her work on FLT, she did work on **Germain's primes**. These are primes p, for which $2p + 1$ is also prime. Examples of these are 2, 3, 5 and 11. As of 2016, the largest such prime has 388,342 digits (see Wikipedia, Safe and Sophie Germain primes). Because of their large size, Germain primes are valuable in public key cryptography. It remains open whether there are an infinite number of Germain primes.

And I have said nothing about her contribution to philosophy, though it was highly praised by the eminent French philosopher Auguste Compte.

In conclusion I want to say that there were many times in her life that she suffered because she was a woman. I'll list some of them below.

(1) Germain's family initially tried to dissuade her from mathematics and science. A boy at that time would have received more encouragement.
(2) She wasn't allowed to attend the École Polytechnique. Although she was easily able to procure the notes of the courses that she chose, she didn't receive an all-round foundation in mathematics. This meant that at least her early work was more naïve and less mathematically polished than it might otherwise have been. It has been noted more than once that her work often had many errors. Some of these were quite subtle.
(3) Although mathematicians like Lagrange and Gauss were happy to work with her, she was not welcomed into the academic arena and certainly there was never any thought of her being granted a position at any of the French universities.

(4) There were times when her work appeared in other mathematicians' work without recognition.

(5) Despite her groundwork in elasticity, she didn't receive the acknowledgement she deserved in that topic. For example, when the Tour d'Eiffel was built, the names of 72 men who had made contributions to the field were displayed on the building. Germain's name was omitted.

(6) Despite her achievements in mathematics, as well as philosophy, she was ineligible to become a member of the French Academie des Sciences because she was a woman. (In 1962, Marguerite Perey was the first woman elected to the Academy. However, she was only a "corresponding member" who had limited privileges.) See

https://scientificwomen.net/women/perey-marguerite-147.

(7) Germain was even badly treated on her death certificate. The official who wrote her occupation used the words "rentiere-annuitant" (pensioner), rather than "mathematicien" as she deserved.

In the late 18th century/early 19th century it was very difficult for women to have an academic life. They certainly had to have a strong will to even get started. How much better is it now? Certainly there are now many female professors, but their number is still significantly smaller than that of male professors.

13.8 Quadratic Forms

Just as quadratic equations were the starting point of Fermat's algebraic geometry, they were also the starting point of great things in his number theory work. In number theory, a quadratic polynomial in two variables with integer coefficients is called a **quadratic form**. For example, $x^2 + y^2$ is a quadratic form and a typical question about it is: which primes have this form? That is, which primes are sums of two squares? For small primes, it is easy to find the answer (as we already did in Section 9.5, but it won't hurt to see some of it again):

Prime	Sum of two squares?
2	Yes, $2 = 1^2 + 1^2$
3	No
5	Yes, $5 = 2^2 + 1^2$
7	No
11	No
13	Yes, $13 = 3^2 + 2^2$
17	Yes, $17 = 4^2 + 1^2$
19	No
23	No
29	Yes, $29 = 5^2 + 2^2$
31	No
37	Yes, $37 = 6^2 + 1^2$
41	Yes, $41 = 5^2 + 4^2$
43	No
47	No
53	Yes, $53 = 7^2 + 2^2$
59	No
61	Yes, $61 = 6^2 + 5^2$

I mentioned in Section 4.3 that Diophantus apparently knew that primes of the form $4n + 3$ are *not* sums of two squares. This is borne out by the table. But the table also suggests – and Diophantus no doubt noticed this too – that the remaining primes *are* sums of two squares. Leaving aside the only even prime, 2, the sums of two squares seem to be precisely the primes of the form $4n + 1$. Fermat claimed that this is true, but we don't know whether he had a proof.

Fermat's **two-square theorem**, if we may call it that, challenged some of the best mathematicians of later generations. Beginning with Euler in

1749, several proofs were discovered, and the theorem became a kind of test case for new methods in number theory. Lagrange, Gauss, Dedekind, and Minkowski, among others, all presented new proofs of the two-square theorem.

However, there is still no really short proof of the two-square theorem, so here I will confine myself to proving the easy part: that numbers of the form $4n + 3$ are not sums of two squares. This happens because any number is either odd or even, that is, of the form $2m + 1$ or $2m$. Therefore, a square is either

$$(2m + 1)^2 = 4m^2 + 4 + 1 = 4(m^2 + m) + 1 \quad \text{or} \quad (2m)^2 = 4m^2.$$

In the language of congruences I introduced in Section 13.3, a square is congruent to 1 or 0, mod 4. It follows that a sum of two squares is congruent to 0, 1, or 2, but *not* 3, so the sum is not of the form $4n + 3$.

As for the hard part, here is just a hint of where number theory went in pursuit of the two-square theorem.

A crucial development, which began with Euler in 1770, was the introduction of *complex* numbers to factorise quadratic forms. Euler himself factorised $x^2 + 2$ as $(x + \sqrt{-2})(x - \sqrt{-2})$, treating the complex number $\sqrt{-2}$ as an "integer". Amazingly, this makes sense and gives a way to solve the equation $y^3 = x^2 + 2$, thereby proving a claim of Fermat that the only positive integer solution is $x = 5$, $y = 3$.

In the case of sums of two squares the appropriate factorisation is

$$x^2 + y^2 = (x + iy)(x - iy), \quad \text{where} \quad i = \sqrt{-1}.$$

Then if x and y are ordinary integers, $x \pm iy$ are called **Gaussian integers** because Gauss studied them in 1832 and showed that they really behave like integers. In particular, there is a natural concept of **prime** Gaussian integer, and a **unique prime factorisation** theorem quite similar to the ordinary one that goes back to Euclid's *Elements*.

It then turns out that an ordinary prime p is a sum of two squares, $p = x^2 + y^2$, if and only if $x \pm iy$ are prime Gaussian integers, and these **Gaussian primes** turn out to be precisely those for which $x^2 + y^2$ is a prime of the form $4n + 1$. For example, since the prime 37 equals $6^2 + 1^2$ the Gaussian integers $6 \pm i$ are Gaussian primes. I don't claim that finding the Gaussian primes is a lot easier than finding ordinary primes that are sums of two squares, but it is enlightening to see the connection between the two.

It is similarly enlightening to see that primes of the form $x^2 + 2y^2$ correspond to "primes" in the world of "integers" $x \pm y\sqrt{-2}$ discovered by

Euler. In fact, Fermat also conjectured that primes of the form $x^2 + 2y^2 > 2$ are precisely those of the form $8n + 1$ or $8n + 3$, and this can be proved by finding the "primes" of the form $x \pm y\sqrt{-2}$. The idea also works for primes of the form $x^2 + 3y^2 > 3$, which Fermat conjectured to be those of the form $3n + 1$. Here are some examples.

Prime $3n + 1$	In form $x^2 + 3y^2$
7	$2^2 + 3 \times 1^2$
13	$1^2 + 3 \times 2^2$
19	$4^2 + 3 \times 1^2$
31	$2^2 + 3 \times 3^2$
37	$5^2 + 3 \times 2^2$
43	$4^2 + 3 \times 3^2$
61	$7^2 + 3 \times 2^2$

But after this, some mysterious trouble starts. Fermat couldn't find a complete description of primes of the form $x^2 + 5y^2$, and this quadratic form also puzzled Euler and Lagrange. Eventually, the anomalous behaviour of $x^2 + 5y^2$ was traced to an anomaly in the "integers" of the form $x \pm y\sqrt{-5}$. They have "primes" but *not* unique prime factorization! This was a decisive event in the history of algebra, because mathematicians wanted unique prime factorisation so much that they had to rethink the whole idea of "integers" and "primes".

Fermat, may not have known it, but he was passing a baton to a whole team of runners in the race to find the secrets of the primes.

Epilogue

Now I hope that you aren't cheating and have read all the book before you set eyes on this Epilogue. If not, what comes in the next few pages will make absolutely no sense. What I'm trying to do here is to take up some of the bigger ideas of the book and to underline them.

The Relay Race

How did the book get this title? Think about it. In a relay race there is a baton that gets handed on from one member of the team to the next. Everyone in the team makes a contribution to getting to the end. It's like this in maths. One mathematician gets an idea and does what they can with it. Most likely they will not solve the whole problem but they make progress. Sometime later another mathematician sees this, likes the idea, knows about things in that area of the subject and gradually takes the baton to the next step. Now more is known about the topic. This process is repeated until that entire part of mathematics has been sewn up. Well, that's not quite right. While the baton appears to have been put down other things happen.

There are several legs of this relay race in this book. For instance, Fermat was reading some number theory. It was Diophantus' book. Inspired by what he was reading, Fermat suddenly had a thought. He knew about Pythagoras' theorem and wondered whether it could be extended. But how can you extend Pythagoras' theorem? It's complete. It tells us all we want to know about distance in geometry.

What happens if you replace the power of two in Pythagoras' theorem by a power of three or four or more or n? Fermat didn't see anything geometric, but he noticed that he couldn't find any solutions for the examples

he looked at. But he thought he could show that there were never any solutions. So he scribbled a note to that effect in his book. Later he thought
some more about it and he wasn't so sure. He gave the idea away. He went
on to something else. But he had found that there are no whole number
solutions for

$$x^3 + y^3 = z^3 \quad \text{or} \quad x^4 + y^4 = z^4.$$

The next step in the process occurred after his son published all his maths.
The baton was now held by the book. People read it and they couldn't
solve it either. But they realised that you only had to worry about the
power of 4 and prime powers. Later Sophie Germain played around with
the problem and saw how to divide the problem into two parts. This was
quite popular as more people took up the baton of Fermat's problem. Then
someone managed to invent computers. There was a sudden rush of baton
carriers who found no solutions for bigger and bigger prime powers. Despite their efforts though, no one was able to show that there were no solutions for any power bigger than 2. Then Wiles came along and knocked
the general case off. Who will be inspired by Wiles' work? Where will
Fermat's Last Theorem go next?

And that is how results in maths, and lots of other things, develop. By
standing on the shoulder of some giant, someone can see further than the
last giant and, by taking up the baton, knowledge can be carried further.
This book is about maths and there are a lot of giants here standing on
the shoulders of some other previous collection of giants. And there are a
whole lot of batons to be carried further.

It's also worth noting that more than one baton can be carried by the
same person at the same time. Wiles proved FLT as a consequence of his
interest in elliptic curves. Of course, many people can have their hand on
the same baton, which would be somewhat problematic if they were all
confined to the same lane.

Mathematical Developments

In this book we have looked at many mathematical developments, but I
just want to give a reminder of four of these: zero, numbers, algebra, and
proof.

Zero became an important way to say that something wasn't there. So
in the numeral 107, the 1 says we have one hundred, the 0 says we have
no tens, and the 7 says we have seven units. This was much better than

writing 17 and hoping that the context told you what the 1 meant. Gradually zero took its place as another number. So $5 + 0 = 5$ and $5 \times 0 = 0$. Zero was a little *different* from other numbers because you couldn't divide by it, but mathematicians gradually became reconciled to this difference.

Numbers grew by steps too. Early on positive whole numbers were the only ones needed for everyday life. But gradually there was a need for negative numbers, zero, and fractions. With the Pythagoreans, irrationals like $\sqrt{2}$ appeared. The number π had been around for a long time, but people found decimal approximations for it and later saw that it wasn't rational, or even algebraic. More such transcendental numbers gradually appeared. But numbers were to develop in more ways yet.

Algebra was one of the great achievements of mathematics. Looking at it one way, algebra has always existed. Euclid used visible lengths to determine unknown quantities, but they were always handled by ordinary language. Gradually symbols arrived along with ways to add and multiply them until finally there was algebra as we know it.

Now a similar developmental process happens with proof methods. They extend and generalise. For example, you can follow the way mathematical induction proceeds. It started off with a slightly vague approach where the author roughly says that having done that you can go on from here forever. Then there is the step where the vague becomes formalised. But there is also the time when the starting case isn't the first case or a single starting case becomes a set of starting cases. Then induction can be even more complicated, but I avoided that. And mathematicians from Euclid on made induction go down and produced infinite descent.

Mathematicians

In this book, who were they? Up to roughly the middle of the 17th century, they were all pretty much amateurs. Euclid and his Greek friends were more or less academics, but most of the others were rich enough to do maths on the side. This trend continued until possibly the middle of the 19th century. It would have been extremely rare before that for someone from a "poor" family to have suddenly burst on the mathematical scene. In fact, things now may not have changed completely. The general route to mathematical fame is generally via an undergraduate degree, a PhD, and then an academic life at some university. This route may well miss some potentially great mathematicians who need to work to keep the family going. And the same is true for any other discipline.

In the past there were only a small number of women who managed to produce original maths. They persisted despite the efforts of their family and society. Things have changed today, but there is room for more female mathematicians.

How Long?

What do mathematicians do? Anyone who thinks about it likely supposes that mathematicians teach, much like school teachers, except that university lecturers have longer holidays than even school teachers do. Of course they may do some research, but that is not a clear concept. So let me try to give a better idea of what goes on. For a start, I have to say that it is very rarely if ever true that a mathematician has a day like this: see a problem; scribble and think; find the proof; write it up and send it off to a journal so that everybody can see the lovely result.

So how long *does* it take to solve a problem? Somewhere between zero and infinity. Pascal and Fermat knocked off the problem of points in a remarkably short time. It was all done by letter over a few months in 1654. If they had been together in one place, they may well have done it inside a month. The length here stands out because people had worried for a long time about what to do if some gambling had to be stopped in midstream. At the other end of the time spectrum, 350 odd years for Fermat's last theorem (FLT) being solved is a really long time. And Wiles actually spent about seven years on his own before he tidied FLT up. And even then, he didn't get it right at the first go.

Cormorants, Wind and Mathematicians

I was very lucky to have spent a significant part of my life in a place where I wasn't far from a peninsula that was not yet covered with houses. I spent many an hour at the weekend looking at the real natives, the birds and other animals that still had a foothold in which they could thrive.

Let me tell you about the *spotted shags* (Heather and Robertson, 2015), rather nice-looking cormorants that lived on the headland at the end of the peninsula. One thing that they did on windy days was to stand on a particular high spot and jump off and glide out into the wind and over the sea. Then, in a great arc, they let themselves float back to where they started. But why did they do this? The winds came randomly and there appeared to be no connection either with eating or mating. But when the winds came the birds spent ages going off the cliff and back. To me it seemed that they were just playing and enjoying themselves. It looked great fun. I could see no other reason for their use of the wind.

Mathematicians are somewhat like this. What is the value of getting all excited about Pell equations? Most of the people who worked on these, at least up to the 17th century, didn't use them in any way, except perhaps Fermat who used them to tease his colleagues, and certainly they weren't motivated by the need to earn a living. Mathematicians were, and still are, mainly doing it for the fun of it and for the satisfaction of solving the problem. Just like people who try to do Sudoku.

The problem may suggest an attack, or it may not. But mathematicians have steps for their gliding. Let me take you through these steps in Figure 13.4.

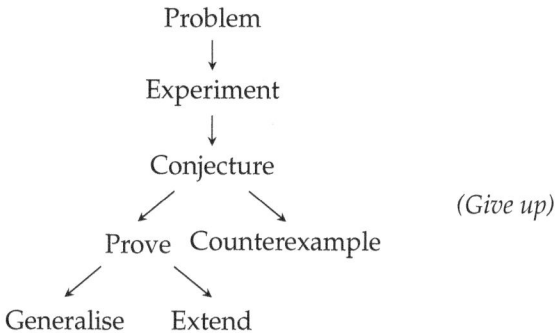

Problem

↓

Experiment

↓

Conjecture

╱   ╲   *(Give up)*

Prove Counterexample

╱   ╲

Generalise Extend

Fig. 13.4 The map of attack to tackle a maths problem

So what is a problem? It's essentially something that you don't know how to solve. I'll give you hints in the other steps on how you might approach solving a problem, but you can't solve a problem until you have one in your hand. So how do mathematicians find problems? Naturally lots of batons with juicy puzzles are waiting to be picked up. You can find problems in books and research papers and conferences and by communicating in the trade. Emailing a friend helps too. Of course, mathematicians make problems up themselves, like Fermat did when reading Diophantus' book. Euclid produced a few too while putting his *Elements* together.

What happens next is to *experiment*. This is the start of play for mathematicians as they produce data and examine it. The process helps them to understand the problem more and gives them ideas of how a solution might be produced. Fermat thought he knew how to solve FLT, but soon found out that his idea didn't work. So he looked at special cases (see $n = 4$ in Section 13.6) to help him see what might happen and how a proof might be developed.

The proof is then a series of steps of logic, mathematical argument, that lead to the desired result when put together. Look at Euclid's proof that the angles in a triangle add to π. After a proof is produced, satisfaction follows. If the problem is significant, then its proof will be accepted for publication in a journal. The conjecture has become a theorem.

At this point there are a couple of things that mathematicians might do. They may think about an **extension** or a **generalisation**. A extension is a variation of a special case that works in a few more cases. Fermat managed to prove his conjecture for $n = 4$. Other people were able to prove it for a few other values of n. When Wiles proved the whole conjecture, he had to prove a generalisation, something that is true for *all* n. Mathematicians are always on the lookout for generalisations because they enlarge the field by not just one more result but by an infinite number. Generalisations may not cover *all* possibilities. For example many theorems about primes hold only "for all odd primes". Such results generalise statements about individual primes because they are true for the infinite class of odd primes.

After a proof a mathematician might think of problems that are like the original problem. For instance, given Euclid's proof that the internal angles of a triangle add up to π, it's tempting to look at quadrilaterals and pentagons and so on to all polygons. Here the work is about extensions. They are small changes to the original problem, changing the shape to see what happens. However, these extensions lead to the generalisation that the sum of the internal angles of every n-gon is $\pi(n - 2)$.

So far I have traced a "good" path down Figure 13.4. But life isn't always that nice. Very often the original conjecture is false. There is a **counterexample**: an example that refutes the conjecture. In that case a new conjecture may come to mind and off you go down the good path again, if it's right! It may still be not quite the conjecture that you need.

One of the most ancient counterexamples is $\sqrt{2}$, whose irrationality (Section 2.6) was a counterexample to the Pythagorean conjecture that all numbers are rational. Counterexamples are sometimes a sign that you have to *give up* – as everyone should if they want to prove that all numbers are rational! But in other cases someone may pick up the baton later.

On the other hand, you may have a sudden new idea. This often comes after a great deal of concentration on the problem. It works both for mathematicians and students doing their homework. I also had an example that happened in an exam! (See Section 4.3.) Lucky me. Mathematicians often go back to a "given up" problem hoping that a new idea will get them home. But a very large number of mathematicians gave up on FLT. So you would be in good company if you gave up on something forever.

It's worth looking for examples of the steps in Figure 13.4 in problems and theorems that you know of. It's also worth talking to others about the steps, to see whether these are basic ideas that apply to other disciplines. For more on this topic, see Holton (2010), Chapter 3.

... to Infinity ... ?

So where does infinity come in all of this? Well if someone has moved a baton further along, mathematics grows larger and mathematicians know more. If we could guarantee that there is going to always be a further step, then mathematics would clearly be infinite. It's like trying to find the biggest natural number. But maybe we need to know more about infinity. Maybe I can take this baton further in a future piece of scribbling?

Bibliography

Adamson, D. (1995). *Blaise Pascal* (St Martin's Press, New York).

Barner, K. (2001). Das Leben Fermats, *Mitt. Dtsch. Math.-Ver.*, 3, pp. 12–26.

Barner, K. (2007). Neues zu Fermats Geburtsdatum, *Mitt. Dtsch. Math.-Ver.* **15**, 1, pp. 12–14.

Bernstein, P. L. (1996). *Against the Gods* (Wiley, New York).

Brieskorn, E. and Knörrer, H. (1981). *Ebene algebraische Kurven* (Birkhäuser Verlag, Basel), English translation: *Plane Algebraic Curves*, by John Stillwell, Birkhäuser Verlag, 1986.

Bucciarelli, L. L. and Dworsky, N. (1980). Sophie Germain. An essay in the history of the theory of elasticity, Studies in the History of Modern Science, 6. Dordrecht-Boston-London: D. Reidel Publishing Company. XI, 147 pages (1980).

Cardano, G. (1545). *Ars magna*, 1968 translation *The great art or the rules of algebra* by T. Richard Witmer, with a foreword by Oystein Ore. The M.I.T. Press, Cambridge, MA-London.

Chemla, K. (ed.) (2012). *The History of Mathematical Proof in Ancient Traditions* (Cambridge: Cambridge University Press).

Colebrooke, H. T. (1817). *Algebra, with Arithmetic and Mensuration, from the Sanscrit of Brahmegupta and Bháscara* (John Murray, London), reprinted by Martin Sandig, Wiesbaden, 1973.

David, F. N. (1998). *Games, Gods and Gambling* (Dover Publications, Inc., Mineola, NY), a history of probability and statistical ideas, reprint of the 1962 original.

Dawson, J. W., Jr. (2015). *Why Prove It Again?* (Springer, Cham), alternative proofs in mathematical practice, with the assistance of Bruce S. Babcock and with a chapter by Steven H. Weintraub.

Del Centina, A. (2008). Unpublished manuscripts of Sophie Germain and a revaluation of her work on Fermat's last theorem, *Arch. Hist. Exact Sci.* **62**, 4, pp. 349–392.

Descartes, R. (1637). *The Geometry of René Descartes. (With a facsimile of the first edition, 1637.)* (Dover Publications Inc., New York, NY), translated by David Eugene Smith and Marcia L. Latham, 1954.

Fibonacci, L. P. (1987). *The Book of Squares* (Academic Press, Inc., Boston, MA), translated from the Latin and with a preface, introduction and commentaries by L. E. Sigler.

Gauss, C. F. (1801). *Disquisitiones arithmeticae*, translated and with a preface by Arthur A. Clarke. Revised by William C. Waterhouse, Cornelius Greither and A. W. Grootendorst and with a preface by Waterhouse, Springer-Verlag, New York, 1986.

Hardy, G. H. (1940). *Ramanujan. Twelve Lectures on Subjects Suggested by his Life and Work* (Cambridge University Press, Cambridge; The Macmillan Company, New York).

Heath, T. L. (1896). *Apollonius of Perga. Treatise on Conic Sections* (Cambridge University Press, Cambridge).

Heath, T. L. (1897). *The Works of Archimedes* (Cambridge University Press, Cambridge), reprinted by Dover, New York, 1953.

Heath, T. L. (1925). *The Thirteen Books of Euclid's Elements* (Cambridge University Press, Cambridge), reprinted by Dover, New York, 1956.

Heath, T. L. (1981). *A History of Greek Mathematics. Vol. I* (Dover Publications, Inc., New York), from Thales to Euclid, Corrected reprint of the 1921 original.

Heather, B. and Robertson, H. (2015). *The Field Guide to the Birds of New Zealand, 4th Edition* (Viking, Auckland).

Hoe, J. (1978). The Jade Mirror of the Four Unknowns – Some Reflections, *Mathematical Chronicle* **7**, pp. 125–156.

Holton, D. (2010). *Problem Solving: The Creative Side of Mathematics* (The Mathematical Association, Leicester, UK).

Hughes, B. (2008). *Fibonacci's De practica geometrie*, Sources and Studies in the History of Mathematics and Physical Sciences (Springer, New York), translated from the Latin, edited and with a commentary by Barnabas Hughes, with a foreword by Frank Swetz.

Katz, V. J., Folkerts, M., Hughes, B., Wagner, R., and Berggren, J. L. (eds.) (2016). *Sourcebook in the Mathematics of Medieval Europe and North Africa* (Princeton, NJ: Princeton University Press).

Katz, V. J. and Parshall, K. H. (2014). *Taming the Unknown* (Princeton University Press, Princeton, NJ), a history of algebra from antiquity to the early 20th century.

Langer, R. E. (1937). René Descartes, *Amer. Math. Monthly* **44**, pp. 495–512.

Laubenbacher, R. and Pengelley, D. (2010). "Voici ce que j'ai trouvé:" Sophie Germain's grand plan to prove Fermat's last theorem, *Historia Math.* **37**, 4, pp. 641–692.

Martzloff, J.-C. (2006). *A History of Chinese Mathematics* (Springer-Verlag, Berlin), with forewords by Jaques Gernet and Jean Dhombres, translated from the 1987 French original by Stephen S. Wilson.

Netz, R. (2022). *A New History of Greek Mathematics* (Cambridge University Press, Cambridge).

Ore, O. (1960). Pascal and the invention of probability theory, *Amer. Math. Monthly* **67**, pp. 409–419.

Plofker, K. (2009). *Mathematics in India* (Princeton University Press, Princeton, NJ).

Russell, B. (1961). *History of Western Philosophy* (Allen & Unwin Ltd., London).

Scott Loomis, E. (1928). *The Pythagorean Proposition.* Cleveland: Masters and Wardens Association of the 22nd Masonic District of the Most Worshipful Grand Lodge of Free and Accepted Masons of Ohio. 214 p. (1928).

Shen, K., Crossley, J. N., and Lun, A. W.-C. (1999). *The Nine Chapters on the Mathematical Art* (Oxford University Press, New York; Science Press Beijing, Beijing), companion and commentary, with forewords by Wentsün Wu and Ho Peng Yoke.

Simmons, G. F. (2007). *Calculus Gems*, MAA Spectrum (Mathematical Association of America, Washington, DC), reprint of the 1992 original, McGraw-Hill, New York.

Smith, D. E. (1959). *A Source Book in Mathematics* (Dover Publications Inc., New York), 2 vols.

Stedall, J. A. (2002). *A Discourse Concerning Algebra* (Oxford University Press, Oxford), English algebra to 1685.

Turnbull, H. W. (1959). *The Correspondence of Isaac Newton, Vol. I: 1661–1675* (Cambridge University Press, New York).

Van Assche, W. (2022). Chebyshev polynomials in the 16th century, *J. Approx. Theory* **279**, pp. 1–16.

Van Brummelen, G. (2009). *The Mathematics of the Heavens and the Earth* (Princeton University Press, Princeton, NJ).

Weil, A. (1984). *Number Theory. An Approach through History, from Hammurapi to Legendre* (Birkhäuser Boston Inc., Boston, MA.).

Index